Praise for Richard Fortey's

Dry Storeroom No. 1

"Fascinating. . . . [An] intimate chronicle of the achievements, the hopes and frustrations, and the virtues and failings of the scientists who strive to bring the museum's collections to life." —*Science News*

"Time amid this collection of fossil collectors, butterfly enthusiasts, eel experts, and amorous botanists is well spent."

—*The Austin Chronicle*

"As he ambles through the museum, Fortey takes no little delight in sticking his head into offices and pulling open drawers containing row after row of specimens. It's an effective way of telling his story: He moves through the museum's different departments, lingering on diversions when it suits him and moving on when he's ready."

—*The Christian Science Monitor*

"Fascinating." —*National Geographic Traveler*

"The role of museums and their importance is the real theme of Fortey's book. And he shows again and again that for all they sound fusty and dusty, they are much more than repositories of the past. They have a vital role to play now and in the future."

—*The New York Times Book Review*

"Enjoyable... will delight a wide swath of readers." —*The Plain Dealer*

"*Dry Storeroom No. 1* lifts the veil on the inner penetralia of the world's premiere natural history museum. . . . [It] is a kind of *The Lives of the Twelve Caesars* of the museum world, mixing the good works, struggles, and scandals of the curators. . . . There is a strong thread running through *Dry Storeroom No. 1* that objects and skills, no matter how obscure, will one day find their use."

—*The New York Review of Books*

RICHARD FORTEY

Dry Storeroom No. 1

Richard Fortey was a senior paleontologist at the Natural History Museum in London. His previous books include the critically acclaimed *Life: A Natural History of the First Four Billion Years of Life on Earth*, shortlisted for the Rhône-Poulenc Prize in 1998; *Trilobite! Eyewitness to Evolution*, shortlisted for the Samuel Johnson Prize in 2001; and *The Hidden Landscape: A Journey into the Geological Past*, which won the Natural World Book of the Year in 1993. He was Collier Professor in the Public Understanding of Science and Technology at the Institute for Advanced Studies at the University of Bristol in 2002. In 2003, he won the Lewis Thomas Prize for Writing About Science from Rockefeller University. He has been a Fellow of the Royal Society since 1997.

Dry Storeroom No. 1

ALSO BY RICHARD FORTEY

The Hidden Landscape

Life: A Natural History of the First Four Billion Years of Life on Earth

Trilobite! Eyewitness to Revolution

Fossils: The Key to the Past

Earth: An Intimate History

Dry Storeroom No. 1

The Secret Life of the Natural History Museum

RICHARD FORTEY

Vintage Books
A Division of Random House, Inc.
New York

FIRST VINTAGE BOOKS EDITION, SEPTEMBER 2009

 Published in the United States by Vintage Books, a division of Random House, Inc., New York. Originally published in Great Britain by Harper Press, an imprint of HarperCollins Publishers, London, and subsequently published in hardcover in the United States by Alfred A. Knopf, a division of Random House, Inc., New York, in 2008.

Portions from Chapter 7 originally appeared in *Orion* magazine.

The Cataloging-in-Publication Data is on file at the Library of Congress.

Vintage ISBN: 978-0-307-27552-3

www.vintagebooks.com

Printed in the United States of America

10 9 8 7 6 5 4 3 2 1

To Leo, with my love

Contents

1. Behind the Galleries 3
2. The Naming of Names 31
3. Old Worlds 73
4. Animalia 114
5. Theatre of Plants 154
6. Multum in parvo 185
7. Museum Rocks 220
8. Noah's Ark in Kensington 257
9. House of the Muses 292

Acknowledgements 315
Further Reading 317
Illustration Credits 319
Index 325

Dry Storeroom No. 1

1

Behind the Galleries

This book is my own storeroom, a personal archive, designed to explain what goes on behind the polished doors in the Natural History Museum. All our lives are collections curated through memory. We pick up recollections and facts and store them, often half forgotten, or tucked away on shelves buried deep in the psyche. Not everything is as blameless as we might like. But the sum total of that deep archive is what makes us who we are. I cannot escape the fact that working for a whole lifetime within the extravagant building in South Kensington has moulded much of my character. By the same token, I also know the place rather better than any outsider. I am in a position to write a natural history of the Natural History Museum, to elucidate its human fauna and explain its ethology. There are histories that deal with the decisions of the mighty, and there are histories that are concerned with the ways of ordinary people. An admirable history of the Natural History Museum as an institution, by William T. Stearn, was published in 1981. What Stearn largely left out was an account of the achievements, hopes and frustrations, virtues and failings of the scientists who occupied the "shop floor"—the social history, if you like. My own *Dry Storeroom No. 1* will curate some of the stories of the people who go to make up a unique place. I believe profoundly in the importance of museums;

I would go as far as to say that you can judge a society by the quality of its museums. But they do not exist as collections alone. In the long term, the lustre of a museum does not depend only on the artefacts or objects it contains—the people who work out of sight are what keeps a museum alive by contributing research to make the collections active, or by applying learning and scholarship to reveal more than was known before about the stored objects. I want to bring those invisible people into the sunlight. From a thousand possible stories I will pick up one or two, just those that happen to have made it into my own collection. Although I describe my particular institution, I dare say it could be a proxy for any other great museum. Perhaps my investigations will even cast a little light on to the museum that makes up our own biography, our character, ourselves.

At first glance the Natural History Museum looks like some kind of cathedral, dominated by towers topped by short spires; these lie at the centre of the building and at its eastern and western corners. Ranks of round-topped Romanesque windows lie on "aisles" connecting the towers which confirm the first impression of a sacred building. Even on a dull day the outside of the Museum shows a pleasing shade of buff, a mass of terracotta tiles, the warmth of which contrasts with the pale stucco of the terraces that line much of the other side of the Cromwell Road. Courses of blue tiles break up the solidity of the façade. The entrance to the Museum is a great rounded repeated arch, flanked by columns, and the front doors are reached by walking up a series of broad steps. Arriving at the Natural History Museum is rather like entering one of the magnificent cathedrals of Europe, such as those at Reims, Chartres or Strasbourg. The visitor almost expects to hear the trilling of an organ, or the sudden pause of a choir in rehearsal. Instead, there is the cacophony of young voices. And where the Gothic cathedral will have a panoply of saints on the tympanum above the door, or maybe carvings of the Flight from Egypt, here instead are motifs of natural history—foliage with sheep, a wolf, a muscled kangaroo.

The main hall still retains the feel of the nave of a great Gothic cathedral, because it is so high and generously vaulted. But now the differences are obvious. High above, where the cathedral might display flying buttresses, there are great arches of steel, not modestly concealed, but rather flaunted for all they are worth. This is a display of the Victorian delight in technology, a celebration of what new engineering techniques could perform in the nineteenth century. Elsewhere in the Natural History Museum, a steel frame is concealed beneath a covering of terracotta tiles that completely smother the surface of the outside and most of the inside of the building; these paint the dominant pale-brown colour. Only in the hall are the bones exposed. This could have created a stark effect but is softened by painted ceiling panels; no angels spread-eagled above, but instead wonderful stylized paintings of plants. It does not take a botanist to recognize some of them: here is a Scots pine (*Pinus sylvestris*),* there is a lemon tree (*Citrus limonum*), but how many Europeans would recognize the cacao plant (*Theobroma cacao*)? Many visitors, and most children, don't even notice these charming ceiling paintings. Their attention is captured by other bones: the enormous *Diplodocus* dinosaur that occupies the centre of the ground floor, heading in osteological splendour towards the door. Its tiny head bears a mouthful of splayed teeth in a grinning welcome.

The *Diplodocus* has been there a long time. It is actually a cast of an original in Pittsburgh, which was assembled in the Museum during 1905. The great philanthropist Andrew Carnegie presented the specimen to King Edward VII, who then handed it over to the Museum in person at a grand public occasion. *Diplodocus* was proudly in place when I first came to the Natural History Museum as a little boy in the 1950s, and it was still there when I retired in 2006. I am always glad to see it; not that I regard a constructed replica of an ancient fossil as an old friend, it is just consoling to pass the time of day with something that changes little in a mutable world.

But *Diplodocus has* changed, albeit rather subtly. When I was a youngster, the enormously long *Diplodocus* tail hung down at the rear end and almost trailed along the floor, its great number of extended vertebrae supported by a series of little props. This arrangement was not popular with the warders, as unscrupulous visitors would occasionally steal the last vertebra from the end of the tail. There was even a box

*I will give the scientific names of all the plants and animals mentioned in this book, because such taxonomy is central to the work of the Natural History Museum.

Diplodocus carnegii, the giant plant-eating dinosaur, with its tail uplifted. It was moved to its present position in the main hall at the end of the 1970s.

of "spares" to make good the work of thieves so that the full backbone was restored by the time the doors opened the following day. Visitors today will see a rather different *Diplodocus:* the tail is elevated like an extended whip held well above the ground, supported on a brass crutch which has been somewhat cruelly compared with those often to be found in the paintings of Salvador Dalí; now the massive beast has an altogether more vigorous stance. The skeleton was remodelled after research indicated that the tail had a function as a counterbalance to the extraordinarily long neck at the opposite end of the body. Far from being a laggard, *Diplodocus* was an active animal, despite the smallness of its brain. Nowadays, all the huge sauropod dinosaurs in films such as *Jurassic Park* show the tail in this active position. Many exhibits in a natural history museum are not permanent in the way that sculptures or portraits are in an art gallery. Bones can be rehung in a more literal way than paintings.

Now animatronic dinosaurs flash their teeth and groan, and carry us back effectively to the Cretaceous period, a hundred million years ago. Small children shelter nervously behind the legs of their parents. "Don't worry," say the parents, "they aren't real." The kids do not always look convinced. The bones that caused such a sensation in Andrew Carnegie's time a century ago, and that still command attention in the main hall, are now sometimes considered a little too tame. There is, to my mind, still something eloquent about the *Diplodocus* specimen: not merely its size, but that it is the assembled evidence for part of a vanished world. All those glamorous animations and movie adventures rely ultimately on the bones. A museum is a place where the visitor can come to examine evidence, as well as to be diverted. Before the exhibitions started to tell stories, that was one of the main functions of a museum, and the evidence was laid out in ranks. There are still galleries in the Natural History Museum displaying minerals, the objects themselves—unadorned but for labels—a kind of museum of a museum, preserved in aspic from the days of such systematic rather than thematic exhibits. Few people now find their way to these galleries.

The public galleries take up much less than half of the space of the Natural History Museum. Tucked away, mostly out of view, there is a warren of corridors, obsolete galleries, offices, libraries and above all, collections. This is the natural habitat of the curator. It is where I have spent a large part of my life—indeed, the Natural History Museum provides a way of life as distinctive as that of a monastery. Most people in the world at large know very little about this unique habitat. This is the world I shall reveal.

I had been a natural historian for as long as I could remember and I had always wanted to work in a museum. When there was a "career day" at my school in west London, I was foolish enough to ask the careers master, "How do you get into a museum?" The other boys chortled and guffawed and cried out, "Through the front door!" But I soon learned that it would not be that easy. Getting "into a museum" as a researcher or curator is a rather arduous business. A first degree must be taken in an appropriate subject, geology in my case, and this in turn followed by a Ph.D. in a speciality close to the area of research in the museum. When I applied for my job in 1970, this was enough, but today the demands are even greater. A researcher must have a "track record," which is a euphemism for lots of published scientific papers—that is, articles on research printed in prestigious scientific journals. He or she must also be described in glowing terms by any number of referees; and, most difficult of all, there must be the prospect of raising funds from the rather small number of public bodies that pay out for research. It is a tall order. Even so, the most important qualification remains what it always was: a fascination and love for natural history. There is no other job quite like it.

The interview for my job was conducted in the Board Room. It was 1970. To reach the rather stern room on the first floor of the Natural History Museum I had passed through several sets of impressive mahogany doors. A large and very polished table was in the middle of the room, the kind of table that is always associated with admonishment. On one wall there was and still is a splendid portrait of the first Director of the Museum, the famous anatomist Sir Richard Owen, by Holman Hunt. He was an old man when he sat for the portrait, and is dressed in a brilliant scarlet robe, beautifully painted to show the glint of satin, indicative of some very superior doctorate. His glittering eyes survey the room, intent on not tolerating fools gladly. Each candidate was interviewed by the Keeper of Palaeontology—who was the head of the appropriate department—and his Deputy Keeper, together with the Museum Secretary, Mr. Coleman. The Secretary was a rather grand personage at that time, who more or less ran the museum from the administrative side. There was also a sleepy-looking gentleman from the Civil Service Commission, who was there for some arcane purpose connected with the fact that the successful candidate would be paid out of the public purse. I was dressed in my best, and indeed only, suit and very nervous.

I was applying to be the "trilobite man" for the Museum. The previous occupant of the post was Bill Dean, who had gone off to join the Geological Survey of Canada. He left behind a formidable reputation. Trilobites are one of the largest and most varied groups of extinct animals, and being paid to study them is one of the greatest privileges in palaeontology. I had not yet completed my Ph.D. thesis, and was young and inexperienced. My fellow candidates were ahead of me by a few months or years. We would all get to know one another well over the course of our professional lives, but for the moment conversation was restricted to twitchy pleasantries. We sat on uncomfortable chairs in a kind of corridor and awaited our turn in the Board Room. Eventually, I had to go in to face the piercing eyes of Sir Richard. The questioning began. Fortunately, I had made some interesting discoveries in the Arctic island of Spitsbergen where I had been carrying out my Ph.D. research at Cambridge University, so once I got going I had a lot to talk about, and my general air of nervousness began to subside. I had discovered all kinds of new trilobites in the Ordovician* age rocks there, and studying these animals seemed a matter of pressing excitement. Youthful enthusiasm can occasionally count for more than mature wisdom. The man from the Civil Service Commission stirred himself once and asked if I played any sport. The answer was no, except for tiddlywinks. He then sank back into apparent torpor. The Keeper smiled at me benignly. Hands were shaken, and it was all over. Did I imagine something less severe in Sir Richard Owen's expression as I left the Board Room?

Several weeks later I was offered the job. In view of my youth I was taken on as a Junior Research Fellow, which meant, I think, that if I did not work out I could be politely escorted out of the cathedral. But important to me was that I was entitled to go behind the mahogany doors into the secret world of the collections, and to receive a modest salary for doing so. I was being paid to do work that I would have done for nothing. I had a season ticket to a world of wonders.

To trace my journey behind the scenes, follow me along one of the few galleries remaining from the old days of the Museum, one flanked by a high wall lined with cases bearing the fossils of ancient marine reptiles: ichthyosaurs and plesiosaurs. They look as if they are swimming along this wall, one above the other, making a kind of Jurassic dolphin pod (although of course they are not biologically related to those similar-looking living mammals). They comprise a famous collection, including some specimens that are the basis of a fossil species name. One of the ichthyosaurs probably died in the process of giving birth to live young, although few visitors notice the label explaining this curious and fascinating fact. Several of the skeletons were dug out by the pio-

*475 million years ago.

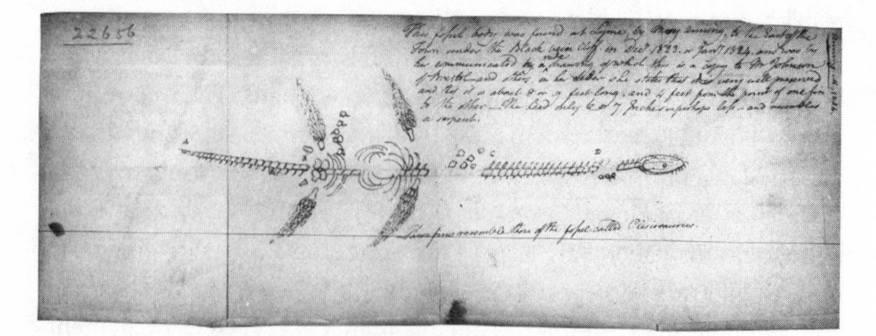

Pen-and-ink drawing of a Jurassic plesiosaur,
made by pioneer fossil collector Mary Anning in 1824

neer fossil collector Mary Anning, who was one of very few women scientists in the first half of the nineteenth century; on summer afternoons an actress may play the part of Miss Anning on the gallery, much to the bemusement of Japanese visitors who think she must be selling something. At the end of the gallery stands the skeleton of a giant sloth from South America, geologically very much younger than the ichthyosaurs. This fine specimen is routinely mistaken for a dinosaur by the more desultory Museum visitors, but it is a mammal, albeit of a special and monstrous kind. Behind the sloth there is a door. And behind the door lies the Department of Palaeontology, home of the really old fossils.

The door opens with a special key. When I first joined the Museum, the keys were issued every day from a key pound staffed by a warder. Every department had a coloured disc attached to the key, a different colour for Botany, Palaeontology, the Office, or whatever. Each member of staff had an individual number. So when I arrived at the key pound in the morning I had to cry out "47 Grey!" and within a few seconds I would be handed my keys by a uniformed warder. When a member of staff became well known to the warder, the arrival of the right keys might anticipate the hollering. The keys were massive, old-fashioned steel affairs such as you might expect to be carried by a "screw" in a prison, or by a miser to open an antique oak chest, and they turned in the locks with a satisfying clunk. There was a specialist locksmith hidden away somewhere in the bowels of the Museum, whose job it was to oil the locks, and keep the keys turning. I soon learned that had I attempted to get into the room where the precious gems were stored I would have discovered that my keys would not fit into that particular lock. There were hierarchies of trust. Presumably only the Director had keys that worked in every lock. We were instructed to keep the keys on our person at all times. Graven into the metalwork were the words "20 shillings reward if found," a measure of the antiquity of the keys, since even in the early 1970s a quid was not much of a reward. From time to time the Secretary would tour the Museum to see which naughty boys and girls had left their keys upon their desks while they went off for a cup of tea, and a ticking-off from above by means of a pompous memorandum would follow. An even worse crime was unwittingly to walk out of the Museum bearing the precious keys. At the end of the working day, the warder could spot a miscreant by an unfilled space in the ranks of keys. Forgetful members of staff were commanded to come back late at night from Brighton or East Grinstead to restore their keys to the hook. A dressing-down would follow from the head of department the following day. The locks were changed in the 1980s to modern Yale varieties, but the new keys were still tailored to different security needs, so I still cannot get to steal the diamonds. By one of those weird volte-faces that only bureaucratic institutions can manage, it is now against the rules to *fail* to take the keys home with you.

Let us go through the doors to the collections. They are housed in a long gallery, across which run banks of cabinets, each some ten yards or so long. There are fifty-seven such banks on the ground floor of the Palaeontology Department, every cabinet neatly sealed by a sliding door designed to keep out the dust. Most of the doors are locked as they are supposed to be. But there is one that has obviously not been sealed away. Carefully slide open the door, and there lies revealed a series of a dozen or so mahogany drawers inside each cabinet. There are labels attached to the middle of the drawers, any one of which might be deeper than the typical cutlery drawer at home. A curator has written a scientific name of an animal in a neat hand on the label, together with some locality information. Pull open the drawer and peer inside: it slides easily on metal runners. There are white cardboard trays on which rest a number of what are evidently bones of various kinds. Even without specialist knowledge it is possible to recognize teeth of several varieties, alongside fragments of limb bones. One of the teeth is a very stout affair, a kind of ribbed washboard on a massive bony base—this is completely characteristic of the elephant family, a monument of masticatory might. These teeth allow elephants to crush tough vegetation of many kinds. All the bones and teeth are more or less stained a yellowish colour. And all of them are fossils, retrieved from the ground by searching strata, digging or scraping in quarries or cliffs; they have acquired the stain of time from their long interment of several hundred thousand years, possibly as a result of the action of iron-rich fluids. Every fragment, no matter how unspectacular it is, tells a story about past time; each one is a talisman for unlocking history. The specimens in this drawer are all fossil mammals, distant cousins of the sloth that guards the entrance to the department.

The collections in this particular part of the Museum and in this particular aisle are devoted to vertebrates from the geologically recent period known as the Pleistocene, a time slice that includes the last ice ages. Inside the tray on which each fossil rests there is a neatly written label which tells us that this particular collection was derived from the cliffs at Easton Bavents, near Southwold in the county of Suffolk, a place where the sea is eroding some of the youngest rocks in Britain, though they are still over a million years old. Sharp-eyed local collectors had spotted these organic remains as winter storms excavated them from the soft sandy cliffs. Had they not been collected and housed in a museum, a few seasons of weathering on those harsh shores would have reduced the bones to meaningless rubble. So the Museum provides a way of cheating decay, of sequestering information from the degradations of time. Doubtless, each specimen provoked a thrill of recognition in its discoverer, the satisfaction of a search rewarded. This single drawer preserves the record of days of endeavour and an archive of pleasure in discovery, or secret gloating over finding the best specimens of the season. Each bone could tell a story of the relative roles of luck and perseverance in science. Fossil fragments have an eloquence that belies their yellowish uniformity. Perhaps the observer will feel a twinge of disappointment at the incompleteness of the specimens, having seen reconstructions in books and films of whole animals striding about the landscape. These remains are just scraps, bits and pieces, odds and ends. The truth is that much fossil material is like this. The skill of the scientist often lies in being able to identify small pieces of a whole animal: from tooth to elephant. Every morsel of the past is useful.

The writing on the labels does not betray any drama of discovery. Old labels like these are written in the hand of the curator at the time the specimen is identified. They are small slips, about the size of one of those special postage stamps issued by countries like San Marino. The writing has to be very neat. Old labels are frequently found written in the copperplate script preferred by the Victorians. Newer ones favour small, neat script. Everything is written in Indian ink so that time will not allow the messages to fade. After all, the 1753 Act of Parliament that set up the British Museum specified that the collections "shall remain and be preserved in the Museum for public use for all posterity." These labels were meant to last. An old label is a message from a curator whom one might never have met, but a little personal message on paper nonetheless. There was a time when the hiring of curators was accompanied by a writing test; nobody with overly large writing would be employed, nor any scribblers, nor any who employed extravagant curlicues. Graphologists would have had a very dull time with those who came through the interviews. More recently, the computer has replaced the skilled human being, as so often, so that neat little labels can be spewed out of a laser printer at the touch of a button. In future, labels will always be impersonal (and if there is a mistake, probably nobody will know who made it). At the top of the label accompanying the large tooth is the Latin, or scientific, name of the animal concerned: *Mammuthus primigenius*—an ancient mammoth. Any visiting scientist will recognize that name. The rock formation from which it was recovered (Easton Bavents Formation) is given next. The age of the specimen within the Pleistocene period follows. Beneath this again is the locality, specified quite precisely. Nowadays a locality might well be given by a GPS position, but British specimens could be fairly precisely located by reference to the national grid, and I have seldom had a problem relocating a locality if this information was given. Then there is the name of the collector of the specimen, who also happened to donate it to the Museum "for all posterity." Many labels will include more information, especially if the specimen to hand has been mentioned or figured in a scientific paper. This is how the importance of the material is conveyed to the outside world: not everybody can come to root around in the drawers of the Museum to see the specimens themselves. Specimens are made known to experts around the world primarily through catalogues and technical publications. So the label might also bear something like: "Figured by Ann T. Quarian in *Transactions of the Society for Ancient Things* Volume 1, Plate 1 figure 2."

That is just one specimen taken at random from a single drawer in a rank of drawers in just one cupboard from one row of cabinets. Some drawers may contain a hundred specimens or more—the next one down includes tiny vole teeth, for example. There may be a dozen or more drawers in a single rank; and there are some ten ranks of drawers in a row. On this floor there are fifty-seven rows or lines of cabinets; except where very large specimens are accommodated, almost every drawer carries a full burden of specimens. In this department alone there are three floors of fossil collections of comparable or greater size. That adds up to a very large complement of drawers, and a vast number of specimens. It does not require a calculation to show that only a tiny fraction of the material held by the Museum is on display to the public: the galleries show the merest sample from a colossal collection. In the secret world behind the scenes there is no shortage of specimens; indeed, one of the main problems is how to accommodate the sheer bulk of new material. Much of it is fragmentary, like the Easton Bavents bones. Its value is scientific and it would not fetch much on the open market. A few specimens are precious and valuable in their own right. "Million dollar fossils" might include the famous original of the Jurassic bird *Archaeopteryx* or the exquisitely preserved fossils of Cretaceous fishes from Brazil. But that is not why we have museums with collections of natural history specimens. A few scraps of bone can tell us what the climate was like three hundred thousand years ago: that is a value that cannot be reckoned in euros or dollars.

My first office was not in the present palaeontology wing, which was officially opened in 1977—by which time I was already an old hand. I originally had an office in the old building, tucked away in the basement beside the main entrance. On busy days I could hear the

Countless specimens: rows of cabinets and drawers for storing the insect collection. In 2007, this storage was being replaced and renewed.

chattering of children as they swarmed up the steps. It was a hugely tall room, and not like an office at all, lit from a large window that looked out on to the lawn in front of the Museum. The collections—my part of the collections—were stored within the room in old storage cabinets. The office was so tall that it had an extra gallery halfway up, reached by a steel staircase. If I wanted to examine some part of the collections, I would have to clunk up the stairs, carrying my hand lens, like an antiquarian gaoler, and open drawers in this upper storey. There were railings all around it to ensure that I did not fall off. The cabinets were beautifully crafted. Each drawer had an independently suspended glass top to keep out the dust. The mortise and tenon joints that formed the corners of the drawers would have struck dumb any carpenter. Labels on the front of each drawer recorded the scientific names of the fossils within. They were cupboards made for eternity. From my first day in that office I felt like an expert—the man from the BM.

I should explain that the Natural History Museum was then known in the scientific trade as the BM, the British Museum. The official title of the museum at the time of my employment was in fact the British Museum (Natural History). The South Kensington museum had split off from the original BM at Bloomsbury when the natural history collections had become so large as to require separate accommodation. The divorce from the mother institution was slow and legalistic. Formal separation from Bloomsbury did not happen until an Act of Parliament of 16 August 1965. The old BM title nonetheless had a magisterial presence that could not be instantly erased. My colleagues would call me up to make a date to "come to the BM" as if that were the only way in which it could be referred to. At conferences, I would still describe myself as belonging to the British Museum—after all, there were other natural history museums all over the place but only one BM, which housed collections made by Sir Joseph Banks and Charles Darwin. However, since the public at large referred to it as the Natural History Museum, in 1990 that finally became its official title. Farewell to the BM, with the finality of the end of the gold sovereign or the landau carriage. Even so, some of my more senior colleagues still sneakily find themselves talking about "finding time to call in at the BM. . . ."

So there I was in my official premises, surrounded by the collections upon which I was to work and to which I was supposed to add. My contract had specified only that I "should undertake work upon the fossil Arthropoda," which left me free to roam through hundreds of millions of years. It might as well have said: "Amuse yourself—for money." But I did have a boss to whom I was accountable. As I have mentioned, the head of department in a British national museum is called the Keeper. This may call up an image of a man in braces mucking out a gorilla cage, or it may have connotations of somebody jangling keys and going around inspecting security locks. It is, however, rather a grand title, one that entitles the bearer to an entry in *Who's Who.* My boss, the Keeper of Palaeontology, was H. W. Ball—Harold William. Above a certain level in the hierarchy one was allowed to call him "Bill"; otherwise, it was always "Dr. Ball." He had the room directly above me, a place of leather-topped desks and filing cabinets. He was guarded by the kind of devoted secretary who exists mostly in the pages of spy novels, like the prim Miss Moneypenny in the James Bond thrillers. She was called Miss Belcher. She was an unmarried lady who lived with and cared for her mother; in the Palaeontology Department she was omniscient. Some years later, I discovered that her Christian name was Phoebe, but I would have no more dreamed of addressing her by that name than I would of addressing the Queen as "Lizzie." She occupied an anteroom through which one had to pass to access the presence of the Keeper; and she always called him that, just as she always called me "Dr. Fortey" until she retired. She regarded such access as a rare and precious commodity, and an audience was a privilege to be awarded reluctantly. In fact, one usually went to see the Keeper because one was summoned. Few employees dropped in for a chinwag.

Occasionally, the summons was for doing something naughty. It was easy to anticipate these occasions. Normally, Dr. Ball gestured towards a chair, beaming, and said something like: "Sit ye down, dear boy." He had a slightly polished-up, satisfied air, like the head boy of a posh school. On the other hand, if you had transgressed one of the rules, you earned a particular stare that P. G. Wodehouse described as "basilisk" when emanating from one of Bertie Wooster's more terrifying aunts. Once I was ticked off for the key offence—leaving them displayed to the world upon my desk. Then there was a diary infringement. The diary was a hangover from the early days of the Museum, being a little book into which the employee was supposed to write his activities, morning and afternoon, and which was collected every month and signed off by the head of department. It was a very tedious bit of bureaucracy, and nobody on the shop floor took it seriously. I took to writing "study trilobites" on the first day of the month and ditto marks for the rest of it. Miss Belcher called me up to say that the Keeper didn't regard this as adequate, and would I please put in more details. So the following month I put in entries like "a.m. open envelopes" and "p.m. post replies" and at the end of the month: "p.m. write diary." My attempts at humour were not appreciated upstairs. The Keeper gave me a flea in my ear and sent me on my way, remarking that nobody was indispensable. Such encounters were, fortunately, infrequent. Diaries were abolished after a few years, and nobody mourned their loss, not even Miss Belcher. The concept of accountability was fairly rudimentary then, so a more usual meeting was an interview once a year with the Keeper to check on my progress. After the "sit ye down" invitation, this grilling usually consisted of noting that I had finished one or two publications that year, jolly good, and see you next year. I had to report on my curatorial assistant, Sam Morris, in similar terms.

Once I was settled into the Museum I vowed to explore the five science departments: Palaeontology, Mineralogy, Zoology, Botany and, in some distant redoubt, Entomology. The hidden museum seemed to stretch in every direction. As more and more new corners were discovered, there seemed no end to it. The public galleries were flanked, underlain and overlain by hidden rooms and galleries and laboratories. There were separate wings and towers. There were odd blind alleys, others that opened into another unsuspected gallery. Some corridors were narrow and poorly lit, and suddenly took a turn downwards into flights of stairs. Others were wider, lined on both sides by mahogany doors carrying the names and titles of the researchers who hid behind them: Dr. J. D. Taylor, Mr. F. Naggs and Miss K. Way were just down the way from my office in the basement. Most of the names were to be matched with faces over the coming months. There were a few I never met face-to-face. Down here in the vaults, there is none of the grand decoration of the public galleries; plain slab floors are the rule, pipes and cable housing run here and there, and almost everything is smothered in institutional cream paintwork. On all sides there are locked cabinets bearing tantalizing labels: Blattidae; Lucinidae; Phyllograptidae. What could they all mean?

Outside my office loomed stuffed elephants and giraffes covered in tarpaulins, dead exhibits that had once graced the main hall. They were now slightly down-at-heel and neglected, with a few bald bits, and rather sad, like a disused sideshow at a fair. The corridors were sealed off into sections by doors that could be opened using the magic keys. It is said that rats, when learning a new maze, make short dashes from home base to start with, gradually extending their range so that unfamiliar territory becomes familiar. So it was with my exploration of the underground or behind-the-wall labyrinth of the Museum. I was able to probe my way from my office in several directions, and I could usually find my way back again. If I got lost, I could pop out of one of the doors into the nearest public gallery to locate my position. Gradually, the most arcane corners of the Museum yielded their secrets.

Westwards along the basement, I let myself through a heavy door just beyond the dead giraffes. There was a notice on the wall that read "Departmental cock"—I never did find out what that meant. Beyond the door, a corridor stretched away lined with polished cabinets on both sides. I had left the Palaeontology Department and entered Zoology. The cabinets housed shells; thousands upon thousands of shells. This was the mollusc section of the Zoology Department, a place where the lingua franca was shells. The cupboard labelled Lucinidae was just one family among many of clams. Any drawer in the stack housed a dozen different species belonging to that family which might come from anywhere in the world, packets of shells laid out neatly in labelled boxes. Many of us have made desultory collections of shells while pottering on the beach on summer holidays: these collections were like an almost infinite and systematic multiplication of that brief acquisitiveness. Dr. J. D. Taylor and his colleagues occupied the offices whose doors opened between the cabinets. Like my own office, they had windows facing out on to the lawns in front of the Museum, and their offices, too, were lined with collections and books, which gave them a cosy, nest-like quality. I soon got to know John Taylor, Fred Naggs and Kathy Way as

Giraffes' heads stored behind the scenes as part of the zoology collections

the mollusc people, the conchology gang, at home with gastropods and bivalves, squids and slugs, nudibranchs and pteropods. As I write this, they are still working in the same rooms, tucked away in their basement redoubt, John Taylor labouring on his beloved molluscs long after most of his contemporaries have taken to the golf club or the allotment. Downstairs from John Taylor's room there was a collection of octopuses and other soft-bodied animals stored in jars, pickled in alcohol and formalin, dead things all pallid and covered in suckers, slightly threatening, as if they might creep out of their accommodation when no one was looking.

At the end of the corridor a small door led to a narrow, dimly lit staircase. It looked as if nobody had passed this way for years. At the top

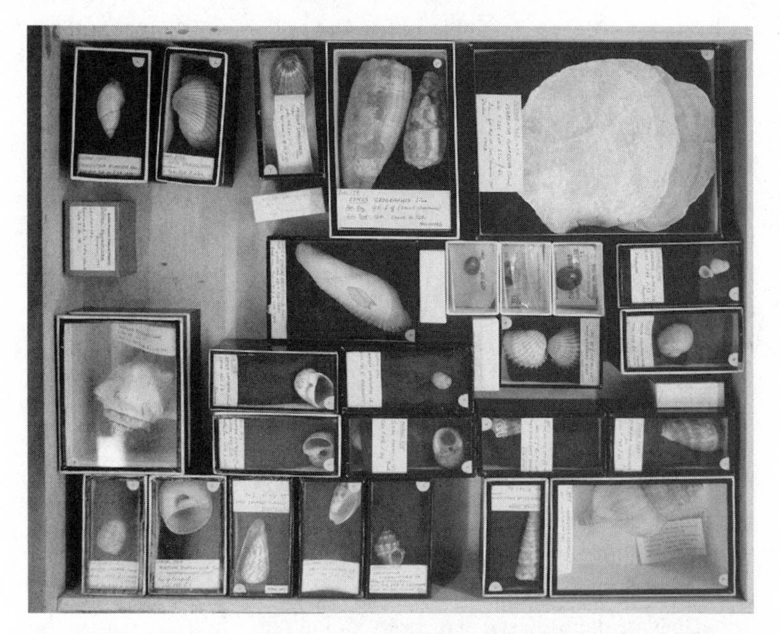

A tray of molluscs from the original Sloane collection, which formed the nucleus of the Natural History Museum

of the stairs was another curious little door, which bore the legend THIS DOOR MUST BE LOCKED. What secrets could be hidden behind it to require such inviolability? It opened out into a broad atrium, and across the way were some huge photographs of insects—beetles, I believe—and a fine formal entrance with double mahogany doors, above which was the notice "Department of Entomology." I had passed through the Zoology Department into the kingdom of the insects. Who could resist the region of the hexapod, the realm of a different Keeper, the habitat of another batch of experts all tucked away from the world in this secret place? Through the doors and beyond there lay another vast empire of the natural world, rank upon rank of cabinets bearing labels identifying the family of insects to which the specimens belonged. I knew that there were further floors above me, and I had a brief vision of swarms of insects beyond number, as in films I had seen of plagues of locusts. Around the perimeter of this huge squarish gallery there were offices with names on doors, Dr. This and Dr. That, all presumably the authorities on the insects in the drawers that lined up in their thousands in ranks in front of me. Perhaps it was not surprising that the drawers themselves were only half as deep as those I had opened in the Palaeontology Department, because insects are mostly rather small, and you can fit a lot into a confined space.

Still, opening one drawer at random, I was surprised to find that there were dozens of butterflies inside, all neatly lined up, as if they were brooches in a jeweller's shop. Every butterfly was pinned tidily through its thorax, with wings spread out to display the fore and aft pairs, each wing shimmering with iridescence as if it had met its death only minutes ago deep in the Amazon rainforest. Some specimens were laid out to show the underneath of the wings instead, which were brown-blotched and mottled, although no less intricate than the dorsal surfaces, if less spectacular. As I pushed the drawer closed again, my gaze wandered along row upon row of similar ranks of drawers. Some part of me tried to do the arithmetic: there must have been about a hundred butterflies in the drawer I had been looking at. Multiply that by the number of drawers in the rank before me, and that number again by the number of cabinets—the mind soon began to reel as the noughts piled on. And to think that the butterfly specimens I happened to be examining were some of the largest and most spectacular of the Class Insecta—the Lepidoptera, the show-offs of the Entomology Department. To be sure, most insects are flies and small beetles, and maybe five times as many of these modest animals could be shunted away inside a single drawer. Many, many more of these insects must have been secreted away on other floors of the department. Hundreds of thousands soon became millions. I need hardly add that very few of these are on display.

My heady calculations were infused with the smell of naphtha, which provided a general fug throughout the Entomology Department. This is a chemical designed to keep away the pests that might otherwise gorge on the insects in the drawers—insects that eat long-dead insects. For of all the members of the animal kingdom the insects are endlessly inventive, experts at survival under almost any conditions, able to prosper where nothing else can earn a living. For some of them, the glue on an old label is a feast. When our own vainglorious species gets its eventual come-uppance—as it will—this will not disturb the cockroach (ah! So here are the Blattidae) one whit, nor jeopardize the prolific weevil, nor distract the swarming aphid. I soon learned that the very success of the insects poses the greatest problems to the entomologist. There are so many species, particularly in the tropics, that simply cataloguing and naming them all can seem an insurmountable task. There is still a long way to go, despite more than two hundred years of descriptive endeavour. We shall see later how scientific ruses have been suggested to get around this labour of Hercules. But, for now, I retreated back down the little hidden staircase into the familiar world of the basement of the Natural History Museum, and to the embrace of the trilobites.

Not far from my office door there is a tiny lift. A brass plaque in the lift informs the passenger that it was installed thanks to the beneficence of Prof. Oldfield Thomas FRS, and it certainly saved the poor curator from walking all around the galleries to get upstairs. I took the lift upwards to the third floor, where the all-purpose key had to be used again to let any passenger out. A kind of cage encloses the elevator's inmates, and as it whizzes away there is an odd sensation of being carried upwards through solid walls. I emerged close by a cross-section through a great giant sequoia tree propped against the wall; this specimen had been displayed to the public ever since the early days of the Museum. I remembered seeing it as a child. Time was spelled out in the tree rings that circled the richly red wood. As evidence of the antiquity of the tree, human events were ticked off along one of its diameters at the appropriate number of annual tree rings. The tree was so big when America was "discovered"; it was of such and such a size when the Black Death stalked through Europe.

There is probably no more graphic way of comprehending earthly time than the stately chronometry of this tree. This one individual plant had seen more than a thousand years of modern human history, yet this was perhaps one-hundredth of the time since our species emerged from Africa. Then again, this life span of *Homo sapiens* (at most a hundred thousand years) was just a late sliver from the great trunk of geological time. The stretch of time life has been on Earth runs to at least 3.5 thousand *million* years. Or, if you prefer, more than a million times the age of the great arboreal Methuselah of living organisms that I was contemplating. Every specimen preserved in the Museum is a product of time, and evolution, cradled in the bosom of our planet. The Natural History Museum is, first and foremost, a celebration of what time has done to life. If the world is to remain in ecological balance, there is a pressing need to know about all the organisms that collaborate to spin the web of life. The planet's very survival might depend upon such knowledge. I want to drag all the visitors to the Museum up to the tree and explain about time, and how we exist atop a vast history that has made us what we are, and that we ignore that history at our peril. But if I did, I fear that I should be branded with the same label as that funny old man who comes up in the street to tell you about his messages from angels.

Not far from the famous tree there is another of those locked doors. By now I knew what to expect. Behind the door there would be a further secret domain, and so it proved. This was the portal to the Herbarium, centre of the Department of Botany. Built almost at the top of the west end of the original Museum, it was the greatest surprise so far. I had become accustomed to the idea that behind the scenes I would find workaday spaces, functional and purposeful, but scarcely matching the grandeur of the public galleries. The Herbarium disabused me of that notion, for before me lay another grand hall, spacious and airy, illuminated naturally from skylights high above. Running along almost the whole length of the "nave" of the hall (which indeed was as long as a large church) there were two ranks of polished hardwood cabinets. By now I knew what to expect of these—they would house the collections. And so it proved. Where a door was ajar I could see folders neatly stacked side by side in the cupboards, a different kind of collection from any I had seen so far. On tables between the cabinets some of the folders lay open for study: each one contained a number of herbarium sheets a couple of feet in length, on which were laid out pressed flowers; well, not just the flowers, but whole plants, leaves, stems and all. The one before me seemed to be a kind of *Aquilegia,* and it was spread out in the most delicate way, so as to display the beauty of its lobed leaves, and the pendent flowers. The fresh green colour of life had faded to a yellowish hue, tinged the colour of dry sherry. But the sheet had preserved the essence of the plant, much as a sepia photograph might preserve a Victorian street scene. There in immaculate copperplate script was the scientific name of the plant recorded by some long-retired curator—the date of collection showed that the plant had been pressed well over a century before, 24 May 1867. These herbarium sheets were clearly as permanent as the other collections I had seen, for all that the "fairest flower [is] no sooner blown but blasted" as Milton said. Death could evidently be stopped in its tracks by using the correct procedures. Then all the archival information could be recorded on the same sheet of paper, not only the name and date, but also the locality, collector and identifier, and details of other specimens in the collection. Once more, I glanced along the rows of cabinets—and there were more on either side of the Herbarium—and tried to guess the vast number of records that must be stored in this great room. Since the herbarium sheet was little more than a slip of paper, there would be dozens of specimens in a folder, many folders to a cabinet, and so on, apparently for ever. The mind soon went dizzy with the calculations. I learned recently that there are more than two million plants stored in the Botany Department.

To either side of the "nave" of the Herbarium there were aisles in which worked the botanists. To be more accurate, the arrangement was like a series of private chapels tucked away on either side of the long hall. There were no doors to separate the workrooms from the main part of the Herbarium, but rather a narrow opening led into a concealed office area, hidden behind the flanking cabinets, a secret niche protected from the prying eyes of the casual visitor. Nowadays, these niches are partly occupied by computers, but when I first visited there were old black typewriters sitting on the desks, and piles of papers and sheaves of carbon paper for making copies for the files. Old monographs lay open at pictures of weeds. The niches did have a Dickensian atmosphere, and one half expected a Sam Weller to pop out from his niche and cry out that he was "wery sorry to keep you waiting, sir." On my exploratory trip into the Herbarium, I was foolish enough to poke my head around the corner of a niche belonging to a very cross-looking senior curator, who threw a kind of generalized snarl in my direction. I decided it was time to flee back to the safety of the trilobites.

Later, I took another trip out through the basement, but this time I popped out of the back of the old building into a kind of alley; this was the "tradesman's entrance" to the Museum, where bicycles were parked, and pallets and unwanted pieces of furniture were piled up. There was often a funny smell (see p. 147). Many pieces of surplus wood or furniture had the words "Rosen wanted" scrawled upon them in white chalk. I subsequently discovered these rejects had been bagged by Brian Rosen, the coral expert. Because his room commanded a view of the alley, he could nip out with his chalk quicker than anybody else. When I got to know Brian better, I visited his house in south London and found complex constructions in his garden entirely fabricated from Museum detritus. There was a kind of apotheosis of the garden shed, a thing with porticoes and pillars, and inside the shed yet more stuff retrieved from the clutches of the skip. I have a vision of a Watts Tower one day rising from the back streets of West Norwood, entirely composed of bits and pieces discarded from the Natural History Museum. Today, the back alley by the bikes is where smokers go to puff furtively, just as they did in their school days.

At the further end of the passageway, an intriguing notice: "Spirit Building" pointed the visitor to a newer block than the original Museum, entirely undistinguished from the outside, no ornament at all. This building proved to be another part of the Zoology Department. The spirits lived in jars. These were the wet collections: pickled, preserved and potted zoology. The storerooms were dark and sealed off from the world, so that when a light was switched on a battery of round glass jars with similar lids stretching away into the distance would suddenly be illuminated . . . and inside the jars maybe a huge python curled up and pallid, like the intestine of a giant, or perhaps a fish all spiny and phantasmagorical, or a lizard that seemed to paw the edge of its jar, or a long-dead lobster. There were several floors of these jars, arranged by zoological kind, cupboard after cupboard full of fishes, or crustaceans, or frogs, or lizards. It was like a storeroom for warlocks, where eye of newt and toe of frog came in a thousand varieties, and fillet of fenny snake was as easy to order as buttered toast.

The specimens were preserved in alcohol or formaldehyde. Colour seldom survived this treatment for long, so that fish and newt, frog and worm, jellyfish and jeroboam, shared a kind of tuberous pallor, something like that of a parsnip. The jars ranged in size from tiny phials crowded together and containing dozens of small shrimps to great towers of glass holding goannas from the outback or carnivorous lizards from Comoro. Here was the profusion of the animal world pickled in perpetuity, a washed-out parade of the panoply of life. It was a place to make one think of the transience of all things. I had never realized until I slid open one of the cupboards in the Spirit Building that many fish have a naturally depressed appearance. A grouper in a jar is a sorry thing indeed, the corners of its mouth turned down in a parody of gloominess. Worst of all is the stonefish, an immensely ugly animal that lurks in estuaries in Australia. It is stout, and covered in warty excrescences, with fins more like props than agents of propulsion; its mouth is gloomier than a grouper's; it seems to have plenty to be gloomy about, looking as it does. I discovered that it spends most of its life camouflaged and motionless, its ragged skin a perfect disguise, until some prey comes near enough—then that apparently dejected mouth can engulf the unfortunate prey in a trice.

On another expedition I encountered the General Library (and this is just one of five libraries in the Museum), after entering it accidentally by a side door off one of the public galleries. It was difficult to believe that there could be so many books pertaining to natural history. Situated in a newer part of the Museum, the main reading room was vast, with tomes on shelves all around the perimeter stretching as high as one could reach—and beyond this room there were galleries of further books, and here they were piled so high on shelf after shelf that there were little ladders to help the reader retrieve some of the higher volumes. There are those who find libraries intimidating, but I am not one of them. I like to see the books in their old leather bindings, the shelves stretching away, deeply filled; it gives me a sense of continuity with past scholars. Even so, encountering an enormous library like this for the first time is a humbling experience. Think of all the thousands of workers putting pen to paper to add to the knowledge of the natural world, or to communicate scientific ideas to their colleagues. If all this is known already, how can a new intruder into the world of learning make any mark at all?

I took down one of the volumes at random: *Acta Universitatis Lundensis*—the scholarly publications of the University of Lund in Sweden.

Well over a century of labours by Scandinavian scientists were preserved here in perpetuity, in volume after volume, or at least until paper crumbles away. The older volumes had green leather bindings, scuffed with use and age; newer volumes were cloth-bound, doubtless in the interests of economy. This part of the library was devoted just to Scandinavian journals, for nearby was a huge run of the organ of the Royal Society of Sweden some yards in length, and over there a great swathe of journals from the University of Uppsala, one of the most ancient universities in Europe. In these pages the great Linnaeus published some of his work, which is still cited today. So it went on, with publications from universities and institutes in Sweden and Norway that meant little to me then. And if the works in this segment of the library were just from one little piece of the world, how much greater would be the literature of the United States, or Russia? Or China? The Museum was dedicated to trying to collect *everything* that was published on the natural history of the planet. Once more I attempted in vain to calculate the size of the holdings on the shelves, floor on floor, only to boggle hopelessly, baffled by bibliographic boundlessness. I crept back to my own little corner.

So my exploration continued, up dark stairwells and down dim passages. I came across a room full of antelope and deer trophies, the walls lined with dozens of ribbed or twisted horns, as if it were the entrance lobby to some stately home owned by a bloodthirsty monomaniac. On another occasion I found my way into one of the towers that flanked the main entrance to the Museum—only to find that to get there one had to take a path that led over the roof. I came across a taxidermist's lair, where a man with an eye patch was reconstructing a badger. I failed to find the Department of Mineralogy altogether, apart from meeting some meteorite experts in their redoubt at the end of the minerals gallery. There seemed to be no end to it. Even now, after more than thirty years of exploration, there are corners I have never visited. It was a place like Mervyn Peake's rambling palace of Gormenghast, labyrinthine and almost endless, where some forgotten specialist might be secreted in a room so hard to find that his very existence might be called into question. I felt that somebody might go quietly mad in a distant compartment and never be called to account. I was to discover that this was no less than the truth.

The geography has changed profoundly since I first entered the British Museum (Natural History). Science departments have been rehoused—my own department, Palaeontology, being the first of them. I had to say goodbye to my polished cabinets, balconies and nineteenth-century haven to relocate to the third floor of the rather characterless modern block tacked on to the eastern end of the old building, an extension as typical of the utilitarian (some might say cheap) 1970s as the old building had been typical of the Victorian love of detail. The relevant minister, Shirley Williams, opened the new wing in 1977. Moreover, the space was generous, and necessary, because the palaeontologists had formerly been dotted all over the place; now they could be together. The Zoology Department, including all those sad fish, has been moved to the much more glamorous Darwin Centre at the western end of the building. Farewell to the Spirit Building, and to its dusty and slightly romantic gloom. Only my old friends the molluscophiles are still secreted in their old haunts in the basement. Nonetheless, the sense of the three-dimensional maze has not been lost. The whole thing just got bigger. Gormenghast lives on.

On one of my forays through the basement I came across a door that I had not noticed before. This was on a corridor with an air of being seldom visited on one side of which were tucked away the osteology collections—bones, dry bones, where oxen strode naked of their skin and muscles, and great bony cradles hung from the ceiling, the jawbones of whales. Here, ape and kangaroo met on equal terms in the demotic of their skeletons, with no place for the airs and graces of the flesh. Strange though these collections might seem, they were as nothing compared with what lay behind the mysterious door opposite. For this was Dry Storeroom No. 1. Neglected and apparently forgotten, this huge square room entombed the most motley collection of desiccated specimens. Fishes in cases were lined up species by species in their stuffed skins; they were presented in faded ranks like a parade that had forgotten the bunting. At one end there was a giant fish that seemed to have been cut off mid-length, such that the posterior part of its body was apparently missing, and it had a silly little mouth out of proportion to its fat body. It was a sunfish, and its cut-off appearance was entirely natural—a faded notice attached to it proclaimed it was the "type." Elsewhere there were odd boxes, one of which contained human remains, laid out in a kind of slatted coffin. The shells of a few giant tortoises hunkered down like geological features on the floor. There were sea urchin shells, and some skins or pelts of things I couldn't identify. Most peculiar of all, on top of a glass-fronted cupboard there was a series of models of human heads. They were arranged left to right, portraying a graded array of racial stereotypes. One did not have to look at them for very long to realize that there was a kind of chain running from a Negroid caricature on one side to a rather idealized Aryan type on the other. This was a remnant of an old exhibit, heaven knows from what era, with more than a sniff of racism about it. Dry Storeroom No. 1 was a kind of miscellaneous repository, a place of institutional amnesia. It was rumoured that it was also the site of trysts, although love in the shadow of the sunfish must have been needy rather than romantic. Certainly, it was a place unlikely to be disturbed until it was dismantled. I could not suppress the thought that the storeroom was like the inside of my head, presenting a physical analogy for the jumbled lumber-room of memory. Not everything there was entirely respectable; but, even if tucked out of sight like suppressed memories, these collections could never be thrown away. This book opens a few cupboards, sifts through a few drawers. A life accumulates a collection: of people, work and perplexities. We are all our own curators.

2

The Naming of Names

In 1976 I almost burnt down the Smithsonian Institution in Washington, D.C. If I had succeeded I imagine that I would now be one of the most famous scientists in the world. Thirty years ago I was a pipe smoker, and the study cubicles in the National Museum of Natural History in the Smithsonian allowed scholars to puff away at their pipes while looking down the microscope. It somehow seemed an appropriate thing to permit. Behind the scenes at the Smithsonian lies a similarly vast labyrinth to that in our Natural History Museum and here, too, are ranks of collections, quietly minding their own business until a visiting expert re-examines them in their labelled drawers. I was interested in looking at trilobite specimens collected years before by one of their staff, E. O. Ulrich. I had my microscope in front of me, and a line of specimens to one hand, while the other picked up the pipe from time to time and waved it about. It was about as contented a scene as could be imagined. At the end of the day I did what pipe smokers do, which is to tap out the dead pipe into the metal waste-paper bin in the corner of the room. The only problem was that the pipe was not completely dead; I had deposited smouldering scraps of Erinmore Mixture among the discarded pieces of paper in the bin. I shut up for the night and went home. Half an hour later, a passing security man noticed billowing smoke arising from my cubicle; fortunately, the fire was quickly extinguished. It might have been a different story had the receptacle been a woven basket. I overheard mumbled gossip about me for the next few days.

Had I destroyed the United States' national collections of natural history the damage would have been irreparable, not just for that country but for the world. The most important single feature of the great collections is that they form the basis for naming the living world. They are the reference system for nature. The curated specimens are the ground truth for the scientific names of animals, plants and minerals. The naming of organisms is called taxonomy, and their arrangement in classificatory order is called systematics. All those cabinets I passed when I explored the Natural History Museum in London were the storehouses for the still-growing catalogue of what is now often termed biodiversity, the richness of the living world. Although it achieves a lot besides, the basic justification of research in the reference collections of natural history museums is taxonomic. So if my Erinmore Mixture had sparked off a major conflagration, much of the basis for the names given to fauna and flora around the world would have gone up in smoke. It would have taken years to sort out. Some museums were bombed during the Second World War, such as the Humboldt Museum in Berlin, and lost many specimens; the resulting taxonomic confusion is still being worked out. That is why museums place such emphasis on the security of their collections; it is an acknowledgement of their permanence: hence the special keys and the security codes.

The taxonomic mission can be briefly stated: to make known all the species on Earth. It sounds easy. This is, of course, only the beginning of finding out about life, because there lies beyond identification so much more, such as how animals live—their ecology—or why there are different species in different parts of the world. Often the taxonomy is inextricably entwined with these other aspects; for example, when two closely related species may be distinguished by having subtly different ecology. European ornithologists will know the instance of the exceedingly similar marsh (*Parus montanus*) and willow (*Parus palustris*) tits, which are readily distinguishable only by their songs. Recognition of species is always the first step in understanding the complex interactions of the biological world. Just as you need a vocabulary before you can speak a language, so it is necessary to have a dictionary of species before you can read the complex book of nature.

Nor is the taxonomic mission as simple as it sounds. There are a few groups of organisms—birds and mammals come to mind—where large size and intensive study over many years means that the vast majority of species have been named and described. Even here there can be surprises, as when a new species of large ungulate mammal was discovered in the jungles of Vietnam a few years ago, or when in 2005 a pristine part of New Guinea was explored for the first time and a rare bird of paradise brought "back from the dead." Bats are turning out to have many more species than was originally thought. I went with a research student into the cloud forest of Ecuador to see a new bat species with the longest tongue in the world that had apparently evolved to feed from (and pollinate) an extraordinary flower with a matching long corolla. It is illustrated in the colour plates. But as a rule there are very few new mammal or bird species to discover, and global concern is rather with conserving those that are known already: by any standard this is a big enough matter.

With other groups of organisms the story is different. The insects are the most obvious example: small, teeming and unnoticed, they fill almost every habitat on Earth. I will choose one example, from a cast of thousands. You have to be a special kind of person to love fungus gnats, but if you look at mushrooms growing in woods you will certainly see these tiny insects flying around the fungi. They are abundant. Most of us encounter these particular creatures as irritating "wormholes" occupied by their larvae that might spoil an otherwise nice-looking field mushroom. But they still fulfil an important function in nature, and they provide a foodstuff for insectivores in their turn. They are a link in the chain. But how many species are there? And how do you tell them apart? Do they feed on lots of different fungi or are there specialists for particular kinds of fungi? All these questions require the attention of a knowledgeable taxonomist, a microscope and skill. Tiny differences in the wings or the hairs on the legs may be crucial in the identification of a species. With luck, expertly identified specimens will finish up as collections in a cabinet marked Family Mycetophilidae ("mushroom

A fungus gnat, *Mycetophila,* perched on a fly agaric

lover") in the Natural History Museum. As to how many species there are, well, at the last count there were 531 different fungus gnats in the United Kingdom alone, and more being added all the time. It is scarcely surprising that there are many more species still remaining to be discovered in the wild. If a new bat can be discovered in Ecuador's cloud forest, it is likely that nobody has even looked at the fungus gnats. And each species will have its own ecological story to tell, another biography to add to the narrative of the natural world. When it comes to status as a species, size doesn't really matter.

At this point one is supposed to put the beetles centre stage, because what is true of the fungus gnats is true of beetles many times over. The geneticist J. B. S. Haldane remarked, when questioned by a cleric about the putative properties of God, that one sure characteristic of the Almighty would be "an inordinate fondness for beetles"; this has become one of biology's well-worn phrases. It is no less appropriate for all its familiarity. One-fifth of all species are probably beetles, most of them leading inconspicuous lives. There are somewhat fewer than half a million species of these animals named so far. When in the 1980s Nigel Stork collected insect species from the tropical forest canopy, he found that many of them had never been collected before—and that many of them were beetles. These were new species, needing a scientific name. It is a massive task just to make an inventory of beetles, let alone understand their biology.

Outside the insects, we can go on to tiny organisms like nematode worms, or smaller still to single-celled organisms—a whole universe of different biological organisations—or down to the smallest of all, the bacteria, where it may be necessary to use biochemistry or genetics to recognize species for what they are. Or we might go down into the depths of the sea where a start to sampling has hardly been made, and every animal is a specialist adapted to eternal darkness and great pressure. Nobody knows how many undiscovered species there are down there, although it is certain that there are many: a single trawl can reveal a dozen new crustaceans. A 2007 report from the Antarctic revealed many extraordinary new crustaceans that had escaped the attention of all marine biologists. There are huge numbers of unknown fungi—tiny ones many of them, or inconspicuous species hidden away under rotting logs. Something like two million species have been named and recognized, and there are certainly an equal number still to name. Many scientists believe that there may be five times that number, considering the habitats that are still so poorly known. There is so much to do, and so few people to do it. An estimate of the number of people who might know about systematics in the whole world came to only about four thousand. And I have not even mentioned fossils: a history of life also needs the myriad characters in the narrative of extinct animals and plants to be identified. So far, there is no indication that we are anywhere near the end of that process, even regarding the largest animal fossils, such as dinosaurs. The labyrinthine anatomy of the Natural History Museum might after all be appropriate to the world it represents, for the realm of nature is truly a castle of Gormenghast, with its half-explored wings and obscure corners where few venture.

Many of the people whose names I had noticed on their offices or cubicles were soldiers in this war against ignorance. They were unseen heroes in a battle against insuperable odds, a battle unnoticed by the million people who pass through the public galleries. A lifetime of endeavour in these cells might be rewarded with the accolade of becoming an "authority"—even a "world authority." This means that the scientist and scholar knows as much about his chosen area as anyone—his views will be sought out on the identification of his organisms by scientists in Minneapolis, Manchester or Mombasa, be they beetles or bats.

The notion of authority is a curious one. It is not something that one says of oneself, so I have never heard anyone introducing him- or herself as "the authority on mushroom gnats." On the other hand it is quite often used to describe a fellow scientist: "Dr. Buggins, the authority on toads" and so on. Just working in a museum like the Musée d'Histoire Naturelle in Paris or the American Museum of Natural History in New York is no guarantee of becoming an authority. This distinctive label is hard earned by publishing and writing on the chosen subject, although it would be difficult to define a kind of critical mass of words when a young scientist passes into an authority. There is certainly no sex discrimination in the title these days, as there would have been a hundred years ago. It is one of those titles that cannot be bought, nor traded, nor given away; it just arrives, like grey hair.

Taxonomic scientists are often referred to by their speciality. Thus "bat man" would be an expert on bats, "worm man" on worms, and an anthropologist would naturally be a "man man." I suppose I was known as "trilobite man," even though it sounds like a creature from a horror film ("worm woman" sounds even worse). Entomologists tend to be even more specific because, as we have noticed, there are just so many insects; a beetle man is, generically, a coleopterist, but he might be a "carabid man" if the Family Carabidae (ground and tiger beetles) was his favourite family. Since there are more than thirty thousand named species of the family, there is plenty to know about this particular group, and it is not hard to imagine how somebody could spend his or her life getting to understand them. The small town I live in has about ten thousand inhabitants, and I am certainly unable to recognize and name more than a small fraction of them. I find it difficult to imagine being on intimate terms with three thousand people, let alone thirty thousand. When a new species is named and described it has to be distinguished from all the others described earlier; and, not surprisingly, mistakes are occasionally made. I once named a trilobite *Opipeuter*, only to find that the same name had been used a year or two earlier for a South American lizard. It is equally unsurprising that taxonomists tend to be obsessive about what they do. You would have to be to remember the details of so many hairs on so many legs.

Nor are they very well paid. "Looking at butterflies all day . . . it's not as if it was real work." This may partly be a relic from the past. There was a time when to be a museum professional was rather a posh occupation. Private schools were a common background among these elite employees, and probably a private income to match. It was the trade of a gentleman. In those days, perhaps sixty years ago, there was a dress code: suits and ties during the week for the scientific staff. In the Natural History Museum, sports jackets were sanctioned for wearing on Fridays only (by the Director Sir Gavin de Beer) because, after all, this was the day when one went down to the country for the weekend. Later, when one didn't, the sports jacket became a kind of signature uniform for the museum scientist, complete with leather elbow patches. It indicated an endearing otherworldliness. Too much smartness might betray the wrong priorities, and an inadequate grasp of the carabids. One of my colleagues always wore a marvellously baggy bit of ancient tweed that looked as if it might house a selection of undiscovered species of small organisms. I hope it was curated somewhere upon his retirement, just in case.

Most of the scientists behind the scenes are there because they are devoted to their organisms, like a good priest to his flock. At the present time, there are fewer scientists who exactly correspond to the specialist as I encountered him when I first walked around the Natural History Museum. As we shall see, there are many different kinds of systematic activities today. But it is still true that the taxonomic mission underlies the research programme. In this respect, a museum is different from a university department, where teaching and research run hand in hand, but research does not have to be related to collections. It is the collections that give a museum its signature, its durability, its ultimate purpose.

The Natural History Museum collections were moved from the mother institution in Bloomsbury; the transfer to the splendid new building was completed by August 1883. Richard Owen, he who had looked down upon me during my interview, was the driving force behind setting up an independent place to house natural history collections. He had previously been Superintendent of the Natural History Departments in the original British Museum, and argued that the collections had become too large for a billet among the antiquaries. Owen was a brilliant scientist and scholar, intensely ambitious, sometimes devious, a British pioneer in the study of comparative anatomy, and a guru of the bones. For example, he named the moa from its skeleton. The moa is an extinct flightless bird that walked around New Zealand until the arrival of mankind almost certainly extinguished the feathered giant; the scientific name was *Dinornis maximus. Ornis* is a bird in Greek, as in the word ornithology; *Dino-* = terrible as in "dinosaur," "terrible lizard"; *maximus* hardly requires explanation. His judgement rarely faltered when it came to appraising what a sample of bones meant in terms of its closest zoological neighbours. Yet he was no evolutionist. He opposed Darwin vigorously, even after the latter's theory of evolution had won the day among the intellectual class in the latter half of the nineteenth century. Owen's vision of a natural history museum was as a kind of paean to the Creator, a magnificent tribute to the glory of His works, a roll call of the splendid species created by His munificence and love for mankind. The words that I used to sing as a child put it thus: "All things bright and beautiful, all creatures great and small / all things wise and wonderful, the Lord God made them all." The cathedral-like entrance to the great new museum, the nave-like main hall, those columns with their decking of leaves or biological swirls, they all had a message. Here was a temple to nature that was also a shrine to the Ancient of Days.

Owen was an establishment figure par excellence. He knew Prime Minister William Gladstone very well in the 1860s, and had even been a tutor in natural history for the royal children at Buckingham Palace. No museum figure of modern times has been so close to the seat of power. Owen knew how to make things happen, and his persistent lobbying eventually yielded dividends in the form of Alfred Waterhouse's vivid new building. Prince Albert, Queen Victoria's beloved husband, was sympathetic to housing natural history collections in his developing cultural "theme park" in Kensington, and opposite the industrial crafts of the "V&A" along Exhibition Road. The Prince's effigy, covered in gold, still broods over the Albert Hall a few minutes' walk north of the Museum on the edge of Kensington Gardens. Owen was trusted to

Richard Owen in old age with the skeleton he helped to reconstruct of the extinct New Zealand moa, the world's largest bird

design a museum with sufficient seriousness to satisfy the Victorian sense of self-improvement through knowledge, or as the Keeper of Mineralogy put it in 1880: "the awakening of an intelligent interest in the mind of the general visitor." Owen certainly intended to display in the main hall what he called an "index museum" of the main designs of animals in nature, intended to be a kind of homage to the fecundity and orderliness of the Creator. However, by 1884, when the Museum formally appointed its first Director, William Flower, the principle of evolutionary descent seemed to be the only acceptable way to organize nature for explanatory purposes. The cathedral had been hijacked for secular ends, and the temple of nature had become a celebration of the power of natural rather than supernatural creativity.

There are large marble statues of both Charles Darwin and his famous public champion, Thomas Henry Huxley, on display in the Natural History Museum. There is also a bronze of Richard Owen. Few visitors seem to notice them, or pause to read their plaques. Darwin and Huxley look out over a refreshment area on the ground floor, so the great men contemplate a clutter of tables rather than the grandeur of nature. A seated Darwin is in the splendour of his old age, every inch the bearded patriarch; Huxley, seated nearby, is brooding and imperious. Richard Owen stands around the corner, in academic dress, halfway up the main flight of stairs facing the main entrance. His hands are slightly outstretched, and at least to my eye there is something clerical about him, as if he were offering a blessing rather than a specimen, although his face is still fierce and commanding. The formality and equality of white stone have somehow ironed out the differences between Darwin and Huxley; it is their enquiring spirit that pervades the Museum. They have become the saints in the place. Oddly, the dark bronze of Owen seems more out of place, as if its metallic heaviness were symbolic of the arguments lost to the presiding genius of Darwin, beatified in marble.

It is curious to reflect that the differences that separated these two men, the bronze and the marble, still count today, well over a century later. London is dotted with memorials to its great scientists. Newton is in the Royal Society; Michael Faraday stands outside the Institute of Electrical Engineers on the Embankment. Yet nobody challenges the insights that Faraday or Newton had into the workings of the world (while recognizing, of course, that understanding has also moved on). Yet there are those who would still side with Owen, against Darwin and Huxley, on the subject of biological evolution—they would seek to reverse their respective historical roles and, no doubt, cast out the marble statues. This view is predicated on the idea that evolution is "just a theory"; and that other theories—which in fact mean only "creation science" or its close relative "intelligent design"—deserve an equal airing. There are some important and interesting matters hidden away in this argument. There are, indeed, some theoretical issues in evolution-

The marble statue of Charles Darwin, in wise old age

ary theory that are still being investigated; indeed, there are whole journals devoted to such questions. Furthermore, this is what science is about—probing questions, not just giving "the answer." Physics and chemistry are no different in this regard—they are full of theories in the process of being tested. So are cosmology and economics. But the crux for the statue of Darwin is a third consideration. The issue of "creation science" is not the kind of theoretical question about kin selection that might be found in a scientific journal, it's about whether evolution happened *at all.* Put bluntly, it is about whether or not we share a common ancestor with a chimpanzee. The descent of all life through evolutionary processes is not a "theory" in the sense that the creationists would have us believe. So overwhelming is the evidence for evolution by descent that one could say that it is as secure as the fact that the Earth goes around the Sun and not the other way. Every new discovery about the genome is consistent with evolution having happened. Whether we find it appealing or not is another question, but personally I like being fourth cousin to a mushroom and having a bonobo as my closest living relative. It makes me feel a real part of the world. So those who promulgate "creation science" are trying to pull off a trick of intellectual legerdemain, a mind jump concealed by jiggery-pokery, mixing in the truly theoretical with what most scientists would simply refer to as the fact of descent. The effect is to try to turn the clock back to a time when immutable versus mutable species was actually a serious debate, a period when Owen and Darwin might have been thought evenly matched for a while. Like Prince Albert, Owen might have finished up gold-plated and Darwin relegated to a back room somewhere in Dry Storeroom No. 1 if only the facts had turned out differently. History has been kinder to Owen than might have been the case. He is recognized as one of the leading anatomists, an outstanding scientific organizer and instigator of a great museum, even if his dreams for it were transmuted.

The hidden rationale behind the displays in any natural history museum I can call to mind is evolutionary, at least as a kind of organizing principle. It does not have to be like that. I can easily imagine an interesting museum in which organisms were arranged by size or colour, or by their utility to mankind. Storage of all the specimens behind the scenes is an entirely different matter, for that has to be systematic. I understand that there is a now a Creation Museum in Kentucky. Its own creators doubtless regard it as a "balance" to all those pesky "evolutionary" museums. It is interesting that the embodiment of respectability for an idea is still a museum, as if a Museum of Falsehoods were a theoretical impossibility. I look forward to a Museum of the Flat Earth, as a counterbalance to all those oblate spheroid enthusiasts.

The Natural History Museum is just one of many in the United Kingdom, and its story could be matched in almost any European country or in North America. The great proliferation of museums in the nineteenth century was a product of the marriage of the exhibition as a way of awakening intelligent interest in the visitor with the growth of collections that was associated with empire and middle-class affluence. Attendance at museums was as much associated with moral improvement as with explanation of the human or natural world. Museums grew up everywhere, as a kind of symbol of seriousness. Universities founded their own reference collections, some of which grew and prospered. Cambridge University has a collection for almost every science department, and Oxford University acquired one of the most beautiful museums of natural history in the world. In some ways, it is a small version of the London museum, but lighter and airier inside, less cathedral, more marketplace. Large towns needed a museum to celebrate their prosperity, and this period of unrivalled industrial growth meant that there were many new fortunes that sought relief in the purchase of collections. Gentlemen needed "cabinets." An interest in natural history was almost as respectable as an interest in slaughtering wild animals. Our mammal collections show that the two interests were far from incompatible, and that an African or Indian "shoot" could easily become a collection. Scotland was redoubtable in the eighteenth and early nineteenth centuries as an intellectual centre, so it is no surprise to find that the Hunterian Museum in Glasgow and the Royal Scottish Museum in Edinburgh both have wonderful collections. Wealthy individuals began to realize that a certain kind of immortality could be ensured by endowing collections in their name, and that this was a rather more tangible result than the prospects in the afterlife. One thinks of the Carnegie Museum in Pittsburgh. The upshot of all this was an explosion in the number of museums paralleling the growth in the numbers of Literary and Philosophical Societies. Nor was this activity confined to the middle classes, as Jonathan Rose has explained in his *Intellectual Life of the British Working Classes.* No, there was general enthusiasm in most social classes for the life of the mind and the excitement of the new exhibit.

And in the case of Britain this enthusiasm was coupled with the expansion of the Empire. The rights and wrongs have been debated, but it cannot be questioned that the British thought they had both a right and a duty to collect, and then collect some more when abroad in the Empire; and then to send the contents of their collections back home—for keeps. In the eighteenth century, the prospect was for plants of "utility and virtue," as Sir Joseph Banks had said. The market potential was very explicitly built into the purpose of collections. Banks' collections made on the Captain Cook voyage in *Endeavour* between 1768 and 1771 were one of the glorious foundation stones on which the Natural History Museum was built. But in the nineteenth century there was an increasing awareness that the study of plants and animals had a value in itself, that indeed there was a *duty* to inventory the glories and variety of the Empire's realms; this was an impulse carried forward into the twentieth century, at least until the liberation of the "colonies." Many of the collectors were amateurs in the best sense—intelligent men and women posted to India or Australia, or another of Britain's many dominions, with time enough to make collections. This was no doubt motivated in some expatriates by the need to alleviate the boredom of duties carried out far from home; others may have made natural history collecting part of a wider programme of exploration. Terrestrial snails were sent to the Natural History Museum by the splendidly named Henry Haversham Godwin-Austen from India, where, among other things, he surveyed the world's second highest mountain, K2, or Mount Godwin-Austen. Many collectors were talented artists, and women, in particular, were often trained in the skills of watercolour painting. Colours in life could be accurately recorded, even if the collecting process dimmed the original. Then, too, the postal system of the Empire was very efficient, so that collectors could receive encouragement and requests for more specimens from the appropriate Keeper or curator at a museum. From some parts of the world, such as Burma, it was easier to communicate then than it is now.

Many private collections made by moneyed individuals eventually found their way into the national collections by bequest or donation. I might mention as one example the collection of Allan Octavian Hume from the Indian Empire acquired in 1885, with 63,000 bird skins, 19,000 eggs and, as an afterthought, 371 mammals. Or there is the famous Lepidoptera (butterfly and moth) collection, comprising some hundred thousand specimens, left as a bequest by Edward Meyrick in 1938, the product of his lifetime's learning and publication. Some collections were purchased; Hugh Cuming's collection of shells was bought in 1866 for the appreciable sum at the time of £6,000, and a special grant was made from Parliament to buy it. It comprised 82,992 specimens. That is more than a dozen for a pound, which actually sounds quite reasonable. Overall, the collections grew apace, mostly filling up those cabinets

Making known the zoological treasures of Empire: the cover of Allan Octavian Hume's journal *Stray Feathers*

behind the scenes, so that the general public would have been unaware of the increase. In terms of sheer numbers, the entomologists always win. According to William T. Stearn, the insect collections had grown from 2,250,000 specimens in 1912 to some 22,500,000 in 1980. For the whole Natural History Museum, the latest figure is eighty million specimens. Such a number is quite incomprehensible as a quantity. Who could say if a huge pile of wheat in front of them comprised a million, ten million or eighty million grains? We simply do not see large numbers that precisely. Perhaps the more meaningful image is that of the ranges of drawers I saw stretching away into the distance when I explored the far reaches of the Entomology Department all those years ago. There was a vision of the scale of life, shelves and shelves of it, stretching away apparently for ever. That is the extent of our responsibilities.

One of the few things one can admire about the British Empire was its propensity to make museums and botanical gardens. A few years ago, I visited the Indian Museum in Calcutta, another huge and serious building in which the artefacts of the subcontinent were housed. The old Indian Geological Survey was nearby, and there were preserved trilobite specimens collected at the end of the nineteenth century, still in their original cardboard boxes. It was as if the whole place had been preserved perfectly since the British left, a kind of fossilized museum. The curation system still worked, although the twenty-first century had yet to impact on the organization. The same old typewriters were still at the deal desks as they had been in 1947. A clerk arrived at 10 a.m. every day and dusted them off with a feather duster, and then proceeded to do little visible work for the rest of the day. Flies buzzed in the sleepy heat of the afternoon. In the same city was a Botanical Garden, magnificent in decay. The old pavilions had literally gone to seed. Once formal flowerbeds were now overrun with native creepers. It was sad to see a system that had once worked so well fallen into desuetude. There were even snakes in the grass. The banyan tree had grown into the biggest in the world, or so it was claimed, with its many branches propped up by the columns of its aerial roots, so that this wholly natural structure looked like the great mosque of the Mesquite in Córdoba. I hope that when India becomes rich on its own account a little money will be spared to restore the Calcutta Botanical Garden. I had seen its well-cared-for equivalents in Christchurch, New Zealand (see colour plate 4), and in Sydney, Australia, both now grown to splendid maturity, and both worthy heirs to Kew Gardens in London. I like the thought of early colonists listing the creation of a botanical garden and museum as one of the earliest necessities, a badge of civilization. One could argue that this particular idea still has currency, even if many of the other colonial ideals have not withstood the scrutiny of history. It certainly indicates that the nineteenth-century administrators took plants and collections seriously. It is of a piece with the growth of museums and civic gardens in nearly every big town in the home country, and with the great exercise in the systematization of nature that prompted all those earnest contributions to the collections of the Natural History Museum in London, and its equivalents around the world. Nature must evidently be known and named, no less than its beauty appreciated.

Perhaps those colonial pioneers dreamed that the biological world would be described before the dawn of the twentieth century. If so, their dream went unfulfilled; that century came and went and still the inventory of nature was far from complete. I have mentioned already how the labour of making all the species known is still in progress in the twenty-first century. It will not be completed at the end of it, because of the sheer size of the task. We shall see below how modern molecular techniques might provide a shorthand way of speeding up identification. The problem now has an added urgency because of the changes, mostly destructive, that mankind is foisting on the environment. Who can predict whether whole ecosystems will be pushed to extinction as a result of global warming? There are so many species out there that have never been named and described, like those I mentioned in the deep sea. *The Times* reported on 27 June 2006 that an average of three new species of animals and/or plants had been discovered in Borneo for every month of the preceding decade—and this in a part of the world where forest has been reduced by 25 per cent since the mid-1980s. Every species on Earth has a biography, and each one is fascinating in its own way. There may be biologies in the deep sea about which we know nothing. Some of them may be useful to mankind in medicine, or in dealing with extreme conditions as we begin to stretch our metaphorical legs to climb to the stars. Who knows? If we allow species to disappear before they have a chance to tell us about themselves, it will be a tragedy to add to the many that our species has already inflicted on the world. The first stage towards understanding is naming—to recognize that this creature before us is different from another already known. I believe we do not have a moral right to imperil the continuation of any species. Who are we, one species among so many, to obliterate the work of millions of years of evolution? Are we like the Greek gods acting on whimsy? Unfortunately, it is difficult to persuade everybody of this moral position. It appears on few political manifestos, except as a kind of harmless truism, vaguely akin to "we must be kind to pretty furry things." It is so much more important than that. I don't want the only record of a species to be on a video archive, or one of those gloomy, pallid faces peering out of a jar in the spirit collections.

Now that it is clear that natural history museums have an increasingly important role in a world whose biodiversity is threatened, I should perhaps explain the nuts and bolts of naming animals and plants. Readers who are gardeners or ornithologists will be accustomed to calling their plants or birds by scientific names. These names provide a common language for all biologists around the world, because they are the official name, the agreed nomenclature. If the name *Larus ridibundus* is used by a Japanese, an American or even an inhabitant of the Philippines, it is the same bird species that is being identified, regardless of the local name; "black-headed gull" just happens to be our British local name for this particular bird, but few Englishmen would know what the Japanese might call it. Different gull species would be just as precisely specified by their scientific names: *Larus argentatus* (our herring gull), *Larus atricilla* (laughing gull to an Australian) and so on. Plants can have many different vernacular names for the same species, even within the same country. In his magisterial *Flora Britannica* Richard Mabey tells us that Cow Parsley is known as Queen Anne's lace, kex, kecksie, mummy die, grandpa's pepper, badman's oatmeal, blackman's tobacco and rabbit meat. *Anthriscus sylvestris* may lack the charm of these local names, but it means the same to all interlocutors, regardless of their origin. The scientific name for a species has to be a unique two words, or binomial, so it differs from human names in this respect, where there is no limit to the number of John Smiths. The name has two parts: first, the genus (or generic) name, which is invariably capitalized; the second, the species (or specific) name, which is never capitalized even if it is obviously named after a person—as in *johnsmithi.* The latter is a convention, as is the italicization of the scientific name, which readily allows recognition of a scientific appellation in a sheet of printed text. When the same generic name appears in a list, it is customary to abbreviate it to the initial letter, as, for example, in remarking that a collection of birds' eggs included examples of those of *Larus ridibundus, L. atricilla* and *L. argentatus.*

If no two animals may have the same scientific name, neither may any two plants. I do not believe it is against the rules to use the same name for a plant and an animal, since there is little chance of confusing an ant with a liana. I have toyed with the idea of naming a trilobite *Chrysanthemum* just to be mischievous. A unit of classification is a taxon (the plural is taxa), and that is why the business of naming them is taxonomy. Scientific names have a long tradition of taking Latin or Greek form. This goes back to the days when scientific communication was in Latin, as the language understood by the intellectual classes across Europe. In the early eighteenth century descriptions and names of plants and animals were often rather unwieldy slabs of Latin. The present simple system of naming and classifying animals and plants was developed in the eighteenth century by the Swede Carl von Linné, who is himself nearly always latinized to Linnaeus: he it was who showed the utility of the binomial to characterize the species of the living world.

Linnaeus' tercentenary was in 2007. As part of the celebrations I was asked to reply to a speech given at the Linnean Society of London by His Imperial Majesty Emperor Akihito of Japan. Thanks to Linnaeus, His Majesty was able to talk to his fellow ichthyologists about his favorite organisms, small fishes called gobies. I was told that the trees in the Imperial Garden are labelled with their scientific names. We all understood one another, and everyone smiled. Linnaeus worked in his maturity in the charming and ancient city of Uppsala; his system triumphed because of its utility and comprehensiveness. He developed his ideas in plant classification as a young man during travels to Lappland—then a daring undertaking. A quirky portrait of him dressed in Lappish robes was actually painted in Amsterdam a few years later, but it does seem that, like Darwin on *The Beagle*, a youthful adventure set him on the course to greatness. His classification of plants was based on such features as counting the number of stamens—it was a *sexual* system. Some young ladies were forbidden to study it because it might bring a blush to their delicate cheeks. Linnaeus' mission to classify knew no bounds: he moved from plants to animals. *Deus crevait, Linnaeus disposuit* (God created, Linnaeus organized) served as his motto. He distributed his binomials far and wide. The Botanical Garden he laid out in Uppsala, with neat beds arranged according to his system, is still in good order. It ought to be one of the holy places for scientists to visit. Even if the simple sexual system has now been superseded, the legacy of the names lives on. Linnaeus' higher and more inclusive levels of organizing organisms into Order and Class and Kingdom are also still used as part of the hierarchy of the system. The labels on the cupboards that I passed in my peripatetic passage around the Natural History Museum were mostly family names, and the family originated as a unit of classification slightly later.* Inside a given cupboard the curator might have placed a number of species belonging to several genera, all embraced by the family whose name is on the door. It is, if you like, a sophisticated filing system, and if you have millions of specimens, the necessity of a filing system that works is patently obvious. I will leave until later in this chapter the question of what the filing system actually *means* in terms of evolution and ancestry, since Linnaeus lived and worked in a pre-Darwinian world, although I should say that like all taxonomists he used the features of the plant or animal concerned as the basis for his

*The full list of zoological ranks from smaller to more inclusive reads as follows (not all of them have to be used): race, subspecies, species, subgenus, genus, tribe, subfamily, family, superfamily, suborder, order, subclass, class, phylum and kingdom (the most inclusive, such as fungi). The botanical system differs in details. Some of the ranks within a kingdom are recognized by similar endings—for example, animal families usually end in *-idae* (the trilobites Calymenidae) and subfamilies in *-inae* (Calymeninae). It is not the same for plants, where most families end in *-aceae* (such as the daisies Asteraceae). The family is probably the most commonly used unit in everyday use: most people, for instance, can tell the cactus family (Cactaceae) from the daisy family.

classification. The convention of using Latin and Greek for names was easy work for the early taxonomists. Most of them had been educated in the classics, and they knew their way around mythology and literature. Quite soon a whole dictionary of gods, goddesses, nymphs and satyrs had been recruited to label the natural world, mostly as generic names. *Daphne* is a flowering shrub, *Daphnia* is a water flea; Daphne herself was a nymph pursued by Apollo, and changed into a bay tree, as always seemed to be happening in those days. The bay tree itself is *Laurus nobilis,* "noble" because the aromatic leaves were used to crown the brows of heroes.

Like *nobilis,* species names often were, and still are, epithets describing some salient feature of the animal or plant in question. A very beautiful plant might be the species *magnifica,* a very ugly one the species *horrida.* The specific names can be much more complicated, produced by splicing several Latin words together, so that a species with bright green leaves might be *viridifolia,* or one with leaves resembling the skin of a crocodile *crocodilifolia;* this complexity is fortunate, since a very large number of names are needed to accommodate all the beetles. It is necessary for the describer to have at least some knowledge of the classical languages because of the rule that genera have gender—masculine, feminine or neuter—and the species name should therefore agree in gender with that of its genus. The suffix on a genus *-us* is masculine and requires a matching *-us* on the species. The suffix *-a* is feminine, so that a commonly cultivated shrub originating from South America is *Fuchsia magellanica* and not *Fuchsia magellanicus; -um* is a neutral ending. Incidentally, *Fuchsia* is named after a famous herbalist, Leonhard Fuchs, who illustrated plants most decoratively two centuries before Linnaeus, and although Fuchs was evidently male, the genus named for him is female. This paradoxical practice is very common in botany: the well-known names *Forsythia, Buddleia* and *Sequoia* are comparable cases. To add a little Gormenghast to the nomenclatural mixture, Fuchsia (not italicized) was a decidedly female character in Mervyn Peake's Gothic extravaganza, thus completing Fuchs' sexual transmutation on the human scale. The epithet *magellanica* is a reference to the occurrence of the shrub as far south as the Straits of Magellan rather than a direct reference to the great explorer. However, as with Fuchs, it is quite common to name a plant or animal genus or species after somebody, often to honour his or her contribution to the field of study. I have done it myself for people who have collected specimens and then presented them to the Museum collections, or for professors who deserve recognition for all their hard work. It confers a modest piece of immortality. In the case of a species one needs to add a genitive suffix—as in *Fuchsia johnsmithi*—to show that this is John Smith's species of *Fuchsia.* There are a few named *forteyi* species of fossil, all of them remarkably handsome examples of their kind. I should add that it is not regarded as good form to name a species after oneself; somebody else has to do it; modesty forbids after all. Nor is it permitted to cause offence by naming a creature *johnsmithi* after John Smith while stating that it is the most unattractive member of the genus. I have to say that Linnaeus himself did not follow this prescription, and named a useless weed *Siegesbeckia* after one of his enemies.

Humour is a delicate matter in nomenclature. The clam genus *Abra* is crying out to be married with the species name *cadabra;* and so it was in a species named by Eames and Wilkins in 1957: *Abra cadabra,* a very satisfactory touch of humour. However, a subsequent authority decided that the species *cadabra* did not, after all, belong in *Abra*—so it was moved to another genus, *Theora,* and there is nothing very entertaining about *Theora cadabra.* This kind of decision happens all the time in systematic work, as a subsequent author concludes from careful study that a given species is better included in a genus different from the one to which it was originally assigned. Effectively, this moves the species from one drawer in the collections to another. Old views are dropped and new combinations of names have to be learned; this process is known as revision.

Almost as good a pun as the *Abra* example is one of the numerous carabid beetles I mentioned above—*Agra phobia.* But my favourite remains the plant bugs described by one G. W. Kirkaldy in 1904. These genera all had the Greek suffix *-chisme,* pronounced "kiss me." Kirkaldy managed to celebrate all the female objects of his affection by adding the appropriate prefix: *Polychisme, Marichisme, Dollichisme* and so on (there were rather a lot of them, apparently). Sexual innuendo is evidently irresistible to some taxonomists. It can be more blatant.

Professor David Siveter of Leicester University is an expert on small crustaceans called ostracodes. In 2003 he and his colleagues published a paper on a magnificently preserved new fossil genus and species from the Silurian of England, which were some 425 million years old, under the resounding name *Colymbosathon ecplecticos.* If I might be forgiven for returning to the territory of "Biggus Dickus," the remarkable fact about this ostracode was the size of its fossilized penis: if we translate the Greek, this Silurian species is "swimmer with astoundingly large penis." Oddly enough, this attracted the attention of the press in a way that few new species have ever done. *The Sun,* always the leader in tastefulness, featured the story under the banner headline "OLD TODGER"; the *Guardian* was hardly less brazen with "Well hung geologist." I doubt whether *Science,* the distinguished magazine that published the original article, has previously been featured in the pages of *The Sun.*

To the scientific name is added the namer: *Abra cadabra* Eames and Wilkins, 1957, or *Colymbosathon ecplecticos* Siveter et al., 2003. This is so that readers will know who first described the organism concerned, and when. It is remarkable how many plants familiar to Europeans were named first by Linnaeus—certainly almost all the common flowering plants. Botanists like their authors to be abbreviated, and Linnaeus is abbreviated to a bald "L."—hence, bloody cranesbill is *Geranium sanguineum* L. The works of Linnaeus are taken as the starting point for all modern scientific names, and everything published earlier is arbitrarily neglected. The beginning of modern nomenclature for plants is his *Species Plantarum* of 1753, and for animals the tenth edition of *Systema Naturae,* 1758. Fungi are different, since Linnaeus did not have much to say about them. The greatest early mycological figure, the "Linnaeus of mushrooms," was another Swede called Elias Fries, who seemed to have an almost uncanny memory for these most fleeting "vegetable productions of nature"; in fact, modern molecular studies have shown that fungi are not really vegetables at all. His great work, published between 1821 and 1832, is a conscious homage to Linnaeus, the *Systema Mycologicum,* and hence mushroom names go back to 1821, although Fries is said to "validate" certain still earlier names, such as that for the familiar fly agaric, *Amanita muscaria,* the archetypal red mushroom

Fossil's willy is pr

The generously endowed fossil ostracode *Colymbosathon ecplecticos* causes a sensation in *The Sun.*

with white "spots," which Linnaeus had already included in his remit. Quite why Sweden, and in particular the University of Uppsala, should have had such a grip of the system of nature is an interesting question. I went to see Linnaeus' farmhouse outside Uppsala at Hammarby to find out if it offered any clues. It is a simple wooden building, now painted maroon, with neat white square windows, no different from a hundred others in the more agricultural part of Sweden—sensible, four square and with a proper feeling for place. Maybe the clue was in the very modesty of the structure; nothing showy, just a monument to hard and consistent work—farmers' virtues, Swedish virtues, Lutheran seriousness.

So far I have said rather a lot about names, but not much about science. The real business of taxonomy is to look closely at the animal or plant in question to assess its features, the business of identification. Only then can you identify a new and unnamed species, or establish whether a previous observer was mistaken about its systematic posi-

THE SUN, Friday, December 5, 2003 41

erved for 425million years

OLD TODGER

HERE is the oldest willy ever found — on a fossil of a tiny sea creature.

The 425million-year-old todger and its shrimp-like owner were discovered buried in rock in Herefordshire.

By JOHN SCOTT

The 3mm creature was sunk in volcanic ash which mineralised, preserving an image of soft tissue like eyes, limbs — and its disproportionately large organ. Experts have called the new species colymbosathon ecplecticos, which is Greek for great swimmer with big whatsit. It belongs to the ostracode family, which still lives in the sea today.

Researcher David Siveter, of Leicester University, has created a hi-tech 3D image to show how similar it is to modern descendants. He said: "The whole animal is amazing."

tion. There is no way of generalizing this process, since every different kind of animal or plant is a distinct proposition. If you are "spider man" you don't climb up walls to save the world as we know it, but you do know a tremendous amount about spider genitalia, because that is the best feature by which to recognize a species. The fern woman will look at the spore capsules on the back of the fronds, and appreciate subtle difference in the way the fronds are subdivided. Flowers and leaves will be the traditional bailiwick of the botanist; spores and microscopic cellular structures on the gill edge will be the province of the fungus man. A crustacean expert will peruse the finest details of the legs and the antennae of his object of study. A mollusc specialist might appraise the colour and ornament of a marine snail, while a lepidopterist will be as familiar with the speckles and dappling of a butterfly wing as he would be with the faces of his own family. One lepidopterist I knew was actually rather more aware of the former than he was of the latter. An ornithologist might listen to songs, spotting their individuality at species or racial level, but then so will an expert on cicadas or bats. Many specialists will take themselves off to the electron microscope, which will afford crisp photographs of the tiniest of organs or ornament on the smallest of animals: bryozoans ("sea mats") stand revealed as decorators as virtuosic as Islamic ceramicists; a tiny mite encrusted with horns and growths as Gothic as an extra in a Dracula movie; the cells of a parasite beautifully embroidered with the equipment they need to carry out their depredations; the teeth—radula—of a mollusc as distinctive as a rack of stalagmites. The palaeontologist will have fewer details at his disposal, and so will be obliged to read as much as he can from the testimony of bones or shells—the wonderfully symmetric test of a sea urchin, the calcite exoskeleton of a trilobite, the tiny pollen grains of a plant that has long vanished from the earth.

The next stage is the library. Although memory is important in identifying specimens, sooner or later it must be checked against the printed record. This is the point where the scientist takes himself or herself off to the journals and monographs, wherein will be found descriptions and synopses of species related to the one under the microscope. The appropriate number of the journal will be found in a catalogue, nowadays on a computer, and then the hunt around the miles of shelves will begin. If a new species is to be named, it is important to check that it has not been described before, no matter how obscure the book or paper in which it might have first appeared. If an author is unlucky enough to miss an earlier name for an organism, then his own will be doomed, for there is an internationally accepted rule that says that the first published name has priority. An unnecessary younger name then disappears into what is termed synonymy. We have already seen that the valid literature goes back into the eighteenth and early nineteenth centuries, so it is not uncommon to find that a species has already been described somewhere else. A great library like that of the Natural History Museum is an enormous asset, because it holds all the old literature. Most university libraries do not. In this regard, systematic science is quite different from physics or chemistry or physiology, subjects in which old literature rapidly becomes obsolete. Most scientists will not cite references dating back more than a decade, and so they will be unfamiliar with the scholarly pleasures of browsing

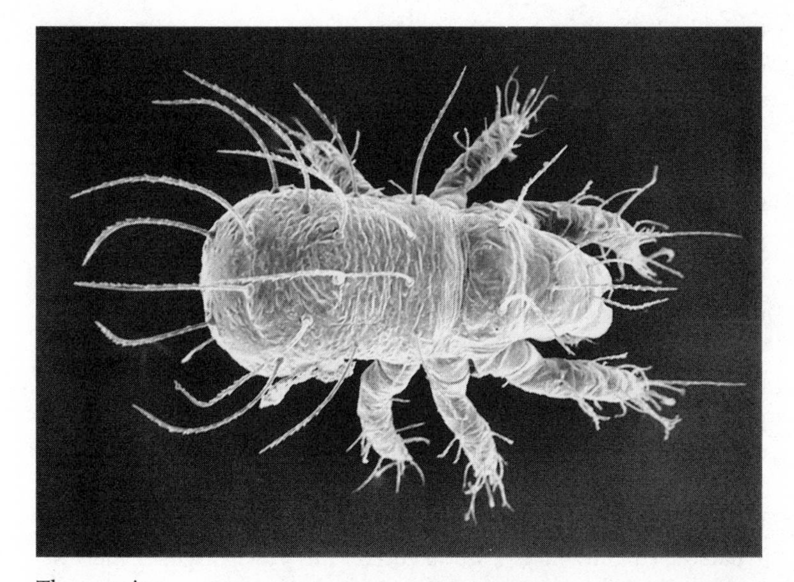

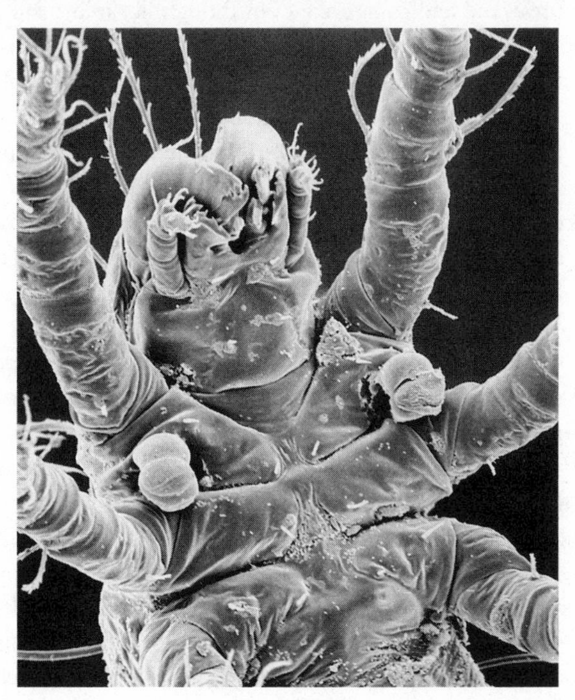

The scanning electron microscope reveals countless unexpected details of taxonomic use. Two views of an orobatid mite larva *Archegozetes* that would fit onto a pinhead: a dorsal view of the whole animal (*above*) and a detail of the head region from below.

through old, leather-bound tomes. It is also a fact that old literature in taxonomy is often as beautifully illustrated as any modern production, particularly the plants, for the drawings of many of the botanical artists of the eighteenth and nineteenth centuries have never been surpassed. Old is not necessarily out of date. Some of my white-coated scientific friends find something amusingly antiquarian about this emphasis on the past, perhaps an image of pince-nez perched on aquiline noses snuffling around in ancient Serbian publications. It is only a little bit true. Most specialists build up personal libraries, and therefore save their legs, and time, in pounding the library floor. The internet has become a wonderful resource for accessing literature, which can now be posted well beyond the confines of the national libraries. There might come a time when all those miles of shelves will be available online from the comfort of home, although I somewhat sentimentally believe that there is an added value in the physical contact with old books. Whatever circumstances arise in future, the paper originals must be preserved and conserved, even though librarians roll their eyes at the sheer quantity of book storage, because cyberspace is not necessarily truthful, and the web can easily become a web of deceit.

It will be problematic indeed to dispense with libraries. At the moment there is a requirement that publications proposing new species should be deposited in one of the copyright libraries—which include our library in London, the Library of Congress in Washington, D.C., and their French, German and Russian equivalents. This is some safeguard against rogue publications and authors setting up new species of animals or plants on spurious grounds. The other safeguard is the system of scientific peer refereeing through which papers submitted to most journals are supposed to pass. An independent reader anonymously says whether the potential publication will pass scientific muster. Neither is foolproof: self-published books can be sent to the libraries and refereeing can be bypassed or inefficient. The eccentric Scottish geologist Archie Lamont set up his own journal, the *Scottish Journal of Science,* which he published from his private cottage in the small village of Carlops. He could just about fulfil the conditions for valid publication, and he set up all kinds of odd-sounding genera of Cambrian trilobites with Scottish names, like *Robroyia* and *Cealgach,* on the basis of miserable scraps. Tails might as well be figured as heads in these works. It has taken years to sort out the taxonomic mess. Almost any other group of organisms will potentially tell a similar tale. Mollusc shells are particularly popular, and the most beautiful of molluscs are unusual snails known as cowries. They show a wonderful and seemingly endless variety of colour patterns, speckled and painted in myriad ways. It is perhaps not surprising that amateur conchologists think they have discovered a new species, and seek the immortality acquired through naming one in publications for their fellow enthusiasts. Many of these claims do not bear close scrutiny, for pigment speckling varies naturally within populations, and not every pattern has a biological reality as a species characteristic. But to sort out the true situation requires all the facilities that a reference library has to offer, ungrateful work much of it, pernickety and irritating. All this labour may eventually be reflected in the small print of a list of synonyms; work at the coalface of taxonomy often lacks glamour.

So now our specialist has carefully looked through the pages of a couple of dozen monographs and papers, comparing illustrations of many species with the specimen in front of him. Piles of old books and reprints of papers litter the office floor. He is convinced that the species he is looking at has never been seen before, based on his wide experience of "his" organisms. It is a new species. He now needs to give it a technical description, illustrate it accurately, give it a new name and then get it published. He thinks that it is an exceptionally wonderful example of its genus, so he decides on the specific name *mirabilis* (Latin, "wonderful," "marvellous"). He checks through all the publications before him; sadly, he finds that a Lithuanian Jesuit has already used the epithet *mirabilis* for a species of the same genus in 1896 in an obscure journal published in Vilnius; this species name is therefore unavailable, and he must find another one. Cursing slightly, he reaches for the Latin dictionary and finds *repanda,* "sought after," instead; good—this one has never been used before, and it will suffice. The next few days are spent in writing an accurate description of the new species, in language as dry as a James Bond martini, with a differential diagnosis saying how it differs from all species known previously. The language is a disguise for the excitement of finding a species new to science, a formal cover-up, or an epistemological stiff upper lip. He might prepare careful drawings under the camera lucida, or supplement his accurate but slightly soulless drawings with photographs prepared by the Museum's skilful studio photographers.

The new species is almost ready to go to publication, but before it can be a valid addition to biodiversity some other important criteria have to be fulfilled. A specimen from the collection has to be selected as the "type specimen"; this is a unique specimen upon which the identity of the new species must ultimately rest. It is known technically as the *holotype* in animal taxonomy, and to be valid must be given an official museum number unique to it. Other specimens in the original collection identified by the author are *paratypes.* Together, these specimens constitute the type collection—the material that provides the material basis for a species' identity in perpetuity: serious stuff. The type specimens are the scientific treasures behind the scenes of the Natural History Museum, a register of biodiversity, held for future generations. They are the ground truth for species in the natural world. Scientists who wish to know whether they are really dealing with the same species will, in the end, have to refer to the types for a definitive opinion. Is this weed that has suddenly taken over crop fields in South America a European invader? Is this fossil ammonite the same as one described in the early nineteenth century from Dorset—and hence are the rocks from which it came likely to be the same age? Is this fly that is plaguing cattle in Namibia the same as one from Libya, and if so how did it get there? Ultimately, the resolution of such questions means that the original specimens have to be examined. Once again, the web is making some difference to how this works out in practice, since it is possible to visit collections in virtual reality. But many fine details—like tiny hairs and microscopic characters—will probably never be accessible over the web. Then there are the sheer numbers involved. A recent estimate puts the London museums' holdings of types at about 670,000; it would be a vast undertaking to put them all online. Originally, the Natural History Museum hung on to its types as firmly as the original BM hangs on to the Elgin Marbles. But now, in more enlightened times, type specimens can travel to recognized sister institutions and bona fide workers. And of course the latter are always welcome as visitors to the vaults. This process probably helps more than anything else in recognizing synonyms, and improving the global standard of taxonomy. So the spoils of Empire have now become a global resource, one that should be recognized by all international bodies concerned with biodiversity.

Scientists deposit their type specimens in the Natural History Museum, or its equivalents elsewhere in the world, because they know that the specimens have been properly curated and cared for there, and should be looked after for future generations. Hence the collection builds steadily in importance as a reference base. There are plenty of examples elsewhere where type specimens have not been recognized for what they were. There are universities that have supported a well-known scholar, and when he or she dies the collections made by the scientist have been assigned to a dusty corner and forgotten. I know of an example where type specimens of fossil ammonites have been rescued from a skip; they might have finished up in the foundations of a building rather than as the foundation of a species. Some type specimens are historically celebrated. The duck-billed platypus (*Ornithorhynchos*) is a bizarre Australian mammal which is famous for laying eggs and having mouthparts like a shoveller duck, not to mention a tail like a beaver. When a specimen was brought to Europe in 1798, it was thought to be a fake, a confection stitched together from different animals by a taxidermist with a perverse or mischievous sense of humour, for it was an animal that should by rights not exist in a well-ordered world. A careful description of the type material proved that the antipodean puzzle really was what it purported to be. We are now quite familiar with its living reality thanks to wildlife photography of the platypus in its natural habitat, where it uses that curious bill to sense small animals on stream bottoms, and the tail to help it swim—not so much an unnatural impossibility as a highly evolved specialist that retains some ancient characteristics. But the type specimen still resides in the collections of the Natural History Museum as a slightly scruffy skin, a veteran of the triumph of science over disbelief. Most types are altogether less famous, and much less conspicuous. Holotypes in the Palaeontology Department are marked only by a modest green spot attached to the rock. Their presence is known only to a small number of specialists and curators. But their importance will not diminish as long as our species pays

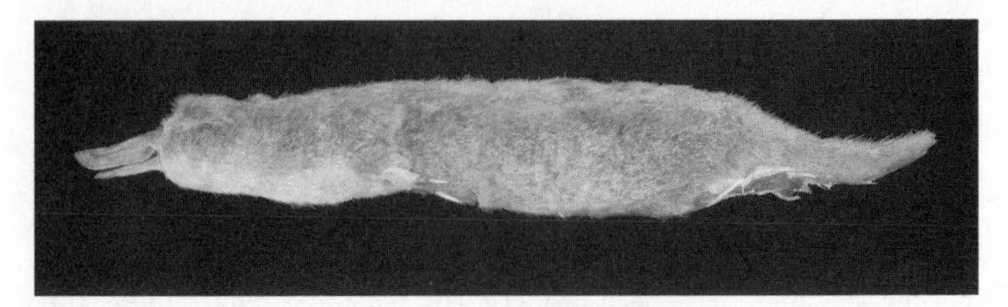

The type specimen of the duck-billed platypus (*Ornithorhynchus anatinus*). This animal was not believed to be real when it was first described.

any attention at all to fellow inhabitants of our planet. The types are still only a small part of the collections; the rest includes comparative material of many more species, or collections made from inaccessible parts of the world, or collections associated with a distinguished individual; so many riches contribute to the archive of the natural world.

The taxonomic process as I have described it would certainly have applied at the time I first nosed my way cautiously around the maze of offices and corridors in the Natural History Museum. I still believe today in the primacy of collections and specimens—they don't go out of fashion, because they are preserved to outlive any passing phase of epistemology. However, it would be surprising if there had *not* been changes in scientific practice and theory over the last decades, if only because science always moves on. I deliberately concentrated on species above, because that basic unit has retained its central role in systematics, no matter how technique and theory have changed elsewhere. Species are not merely specious.

The most important change in the scientific firmament was the appearance of molecular techniques. The possibility of sequencing genes followed upon the unravelling of the structure of DNA—and now has reached new heights after the decoding of entire genomes, including that of our own species. What began as a major technical challenge is now almost entirely routine, and every research institute worth its salt, including the Natural History Museum, has a molecular biology laboratory, staffed by scientists of the white-coated variety, slaving away with test tubes in front of highly sterile machines. Nowadays, an organism must reveal its secrets down to the molecules in its DNA or RNA. Gene sequences provide a whole plethora of characters to add to the traditional morphology—something to challenge the hairs on legs, spines on shells, pattern of bones or structure of flowers. Because the genome is almost unimaginably huge, the potential for information locked in its sequences of bases is theoretically almost endless. It is small wonder that there has been a boom in the employment of molecular biologists at the expense of traditional experts on groups of organisms.

More than twenty years on from the appearance of these techniques it is possible to see just how many questions can now be tackled which were previously beyond reach. Many people have used the obvious pun "designer genes" before, but it is not a bad phrase to summarize what scientists actually do with the vastness of the genome. They use different parts of it for different purposes. If they have been curated appropriately, pieces of type specimens can even be fed into the DNA factory, thanks to a technique known as PCR that "magnifies" sequence information from tiny pieces of tissue. There is, of course, much variation in the genome within a species. Some variation is at the level of the individual—hence the possibility of "nailing" a criminal for an offence using stored samples such as blood or semen years after a deed has been committed. The gene sequences in question identify a particular person beyond doubt, like a fingerprint. Other changes in gene sequences are conserved for slightly longer periods of time; sections of DNA called microsatellites have high rates of mutation, which makes them ideal for studies within the historical time span and within species—for example, in tracing movements of human populations around the world. Other parts of the genome change still more slowly, and yield sequences that are of particular use in recognizing species—we will come back to these again, because they are of special importance in taxonomy. Other parts of the genome are generally conserved, which means they accumulate changes only very slowly, over millions of years, or even longer. Some of these genes are important in the functioning of any organism—they include genes that encode proteins, for example. Or there is the RNA of the cell's "powerhouse" organelle, the mitochondrion,

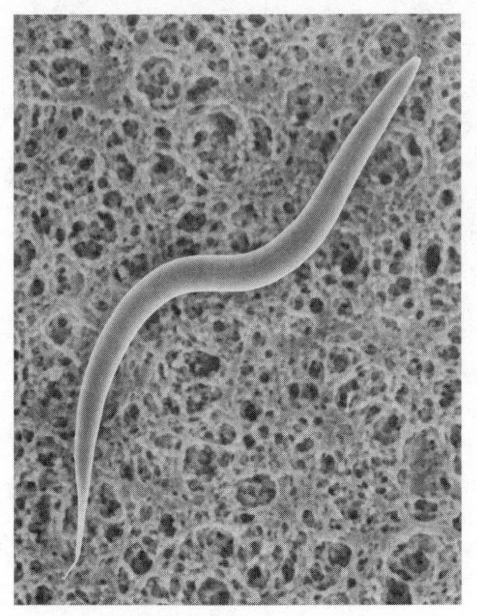

The small nematode worm *Coenorhabditis elegans*— so important in working out the genetics of all animals

which was one of the first molecules of this kind to be completely sequenced. Such slowly changing genes and sequences allow the scientist to "see" backwards in time to the divergence of major lines of evolution, to examine relationships between different groups of organisms that might previously have been investigated only by the palaeontologist delving deep in the fossil record. To say that these discoveries had a profound effect on systematics would be a considerable understatement: they provided a whole new way of looking at the natural world. There are even genes that could potentially "see" the separation of the major designs of animals and plants hundreds—even thousands—of millions of years ago. In 1991 great surprise greeted the discovery that the sequence of the elongation factor gene in the nematode worm *Coenorhabditis elegans* was more than 80 per cent similar to that in a mammal; here was common ancestry writ large. Some genes were evidently so deep-seated that they continued to do their work over a timescale of many, many millions of years. Such evidence proved beyond question that we are one with the worm and the bacterium.

The author handles an edible black true truffle (*Tuber*) from Sardinia.

Evidence from molecules was quite quickly incorporated into the intellectual armoury of the more forward-thinking systematists. For a while there was resistance in some quarters by experts who trusted implicitly their traditional characteristics for classification—colour, or hairs on legs, or behavioural patterns—and did not like the suggestions of new evolutionary relationships thrown up by molecular studies; and it was also true that in the early days some dubious conclusions were drawn from using the wrong "designer gene" for a particular job. However, it was soon recognized that sequencing evidence could provide answers to questions that had been troubling systematists for years. I will give just one example. Edible truffles are subterranean fungi, belonging to the genus *Tuber*. There are several species, and gourmets dispute their relative merits. *Tuber magnatum,* the white truffle, which grows in Italy, commands the highest prices—up to about $5,000 a kilogram. It is the most expensive foodstuff in the world. The Périgord truffle, *T. melanospermum,* is mostly French in origin, and black rather than white. The warty summer truffle, *T. aestivum,* grows in England, but is less sought after, although it is the only one I have found in the wild. All are remarkable for having an extraordinary, and some would say irresistible, odour, which suggests a kind of mushroom/meat hybrid. This intense fragrance is imparted to oil or eggs, and indeed the simplest way to eat truffles is to use them to flavour an omelette, or to grate them finely over scrambled eggs. *Pâté de fois truffé* is such stuff as gourmet dreams are made of. The edible properties of the truffle are not matched by their aesthetic ones, for most truffles look like some kind of knobbly animal excreta which have been passed with not a little discomfort. They do not have to impress with their appearance, for it is the smell that matters. In the wild they grow close to the roots of trees, particularly oak (*Quercus*) and hazel (*Corylus*); they are one of a very large number of fungi that form a symbiotic relationship with the tree host, their mycelium enveloping or penetrating the roots in a so-called mycorrhiza. The host benefits from ions such as phosphate that the fungus can "hunt" from the surrounding soil, while the fungus receives products of photosynthesis from its tree host in return. The problem for the fungus is how to spread its spores from underground and hence ensure its survival, and this is where the smell comes in. The spores are enclosed in chambers or fissures inside the truffle. Animals such as wild pigs find the smell of the truffle attractive, and will greedily grub up the fruit bodies. "Truffle pigs" are trained to smell out the subterranean booty, which is removed from them before they can gobble it up. But when ingested, the spores will eventually pass out of the animal, unharmed, in droppings, having by then been dispersed widely from their point of origin. In rainforests in south-eastern Australia I have seen holes scratched by marsupial potoroos in search of truffles—very different creatures performing the same favour for a truffle on the other side of the world.

When the truffles were first recognized as fungi rather than some spontaneously generated freak of nature, it was thought that such curious productions comprised a single group of organisms—a reasonable assumption, one might think. They deserved one of Linnaeus' high-level classification tags—an Order. But when microscopes came to be focussed on the tissues inside the truffle, where the spores were developing, an interesting discovery was made. Not all truffles were alike. Those that graced the tables of the rich and hedonistic showed features at the microscopic level like those of another gourmet treat, the morel (*Morchella esculenta*). In other, and more technical, words they were ascomycetes. These fungi bear their spores inside minute sacs or asci of the order of a tenth of a millimetre long—there are usually eight such spores, so the asci have a very typical microscopic appearance, rather like eggs wrapped in a sausage. However, some other truffles, for example a genus called *Hysterangium*, showed evidence that they were related instead to the gasteromycetes—the group of fungi that includes puffballs and stinkhorns. These are basidiomycetes, which carry their spores in an entirely different way from the ascomycetes; they are typically borne atop a special cell called a basidium, usually four spores in a loose cluster. The white mushrooms that fill vats in supermarkets are distantly related basidiomycetes, as are the majority of fungi that troop through the woods in autumn. The ascomycetes separated from the basidiomycetes very early in earth history, and certainly more than a billion years ago. It is preposterous to classify truffles together that have such different evolutionary origins—and so the ascus-bearing truffles were separated from the basidium-bearing truffles: so far, so sensible, and resulting in two Orders. For common names we now had "truffles" and "false truffles."

However, the story did not end there. From other microscopic hints there were suspicions that there were several origins for truffles in *both* the ascomycete group and the basidiomycete group. Truffles might have arisen repeatedly, on separate evolutionary trees, for all their superficial similarity. The closest relatives of a truffle might prove to be one of several different kinds of more normal-looking mushrooms and other fungi. The truffle shape, including its subterranean growth, is a specific adaptation—a mode of life, if you like. It was not so difficult to imagine a "truffle habit" originating several times, because most fungi do indeed develop underground, and only later erupt at the surface. If development were somehow "arrested" at the early stage—well, then you might have something like a truffle. The trouble is, how could you pair the truffle with its closest-related mushroom, since there is so little general resemblance between them? This is where the molecular evidence should come into its own. The appropriate mushroom partner should, in principle, show more similar sequence patterns at the molecular level to its truffle relatives than it does to other truffles or indeed other mushrooms. So it has proved. Using the appropriate genomic tool, especially one known as ribosomal ITS (Internal Transcribed Spacer), the complexity of the origin of truffles has been demonstrated. It turns out that at least six different kinds of mushrooms—that is, the basidium-bearing kind—have become "truffleized," to coin a term. To add to this there are several more origins of truffles of the ascus-bearing kind, of which the true truffle, *Tuber,* is one. Far from being a natural group of organisms, the truffles originated from numerous different fungi on several different occasions, and it all probably happened millions of years ago.

Why should anyone care about such apparently esoteric information? After all, most people can happily pass their lives without seeing a truffle of any kind, and who but an outstanding eccentric would spend hours carefully digging around in the litter under trees to find false truffles of the inedible kind? But then, who would guess that truffle evolution was crucial to the survival of several charming Australian marsupials? For the Australian group of truffles, including some placed in the genus *Hydnangium,* were also independently evolved in close association with *Eucalyptus* trees. These false truffles provide a prime foodstuff for bettongs and potoroos, which are delightful, nocturnal cat-sized animals that are now the focus of intensive conservation efforts. The more that is known of their requirements, the more likely they are to survive in the twenty-first century. False truffles are as important to their continued existence as keeping them from the depredations of feral cats. So what might at first seem extraordinarily specialized information has links to those "pretty furry things" after all; nature is seamless, its connections multifarious.

The truffle example also links back to where we started—the questions of taxonomy. Every time a truffle under examination turns out to be related to an entirely different mushroom, we can imagine a curator cursing quietly under his breath and moving the relevant preserved specimens to a different drawer. This is an extreme case of "revision"—revisiting taxonomy. The point is that we expect classification systems, genera, families and so on in ascending order, to reflect fundamental resemblances between the species included in them. The species themselves are the units of this classification—at least they are if we have recognized them correctly—and they are the real things that get shifted around from one drawer to another. The genus or family whose name might be written on the drawer or cupboard is a *theoretical* concept, subject to change as science advances. As with the truffles, species may be added or taken away or moved around. The up-to-date taxonomist wants his classification concepts to square with modern views. For most such scientists this means that the species included in a genus, for example, should have descended from a common ancestor—that is, constitute what is known as a *clade.* The characters shared by the species in a genus—and nowadays these can be molecular characters as much as the traditional "hairs on legs"—are what define it, make it a natural entity. Discover new characters and the concept of the genus may well change, and so will the species included within it. This results in changes in generic names for a given species that irritate many people, and particularly knowledgeable amateur scientists. "Why do they have to keep changing the names?" is a common complaint. However, the contemporary investigator is obliged to seek out genera, or families, that are clades; the scientific method used in recognizing these groups is known as cladistics; and the whole business of examining relationships between organisms in this way is usually termed phylogenetic analysis, or simply phylogenetics. If names have to change as a result of careful reconsideration of species, well, that's the price of progress. Much modern taxonomy is based upon computer analysis of relationships, where all the characters possessed by a group of organisms under study are allowed to fight it out until the "best" arrangement of species is discovered, resulting in a diagram—a cladogram—showing how species relate to one another. The eventual classification is then drawn up directly from the cladogram. For example, several clades of species clustering together might be recognized as separate genera, and if these genera then cluster together in a more inclusive group this larger group might be the basis of a family.

This sounds technical, and so it is. Quite a few famous taxonomists are computer experts first, and lovers of organisms second. They think in algorithms rather than algae. They are mostly interested in animals and plants as experimental material for their classificatory computer programs. Their conversation tends to revolve around the statistical criteria for the support of one piece of the cladogram or another; an outsider hearing these people chatting might think she was overhearing an unknown Amazonian language. However, arcane though it might sound, the cladistic approach has made taxonomy much more of a science, and less dependent on the word of an authority alone. It provides a unifying method across the spectrum of organisms, from virus to vicuña, and can embrace all kinds of evidence, from the molecular to the anatomy of a blue whale. But it will be clear by now that it also makes problems for that Linnaean system of naming animals and plants. Linnaeus himself designed his "system of nature" before the notion of evolution had gained currency. Some might have considered that the order of nature might be an expression of the mind of God alone: "he made them high and lowly, he ordered their estate," as the hymn puts it. The idea that classification might involve notions of descent from a common ancestor was a subsequent introduction. The species as the unit of currency of classification was the only thing in common between these pre- and post-Darwinian worlds. And with the arrival of cladistics and molecular analysis the old Linnaean system might be seen to creak and groan under the stress of frequent changes in nomenclature—so much so that some scientists have tried to persuade their colleagues that the time has come to abandon the Linnaean binomial altogether. They want to replace it, or at least augment it, with something called the PhyloCode.

As this is written, the PhyloCode is still undergoing its own evolution, and it might be premature to anticipate the outcome. Many critiques of the Linnaean system are surely correct. There is no consistency in the use of the ranks of the system between different kinds of organism; some parts of the natural world have small genera, other parts have large ones, and a family can be a very different concept from one worker to another. We already have an intuitive feel for this. Birds are finely divided into genera separated by tiny anatomical differences; on the other hand, some genera of plants and fungi might include several hundred species. The attractive sea snail genus *Conus* includes at least six hundred species. The recognition of what makes a genus or family is partly a matter of tradition and taste. It is also undoubtedly true that there are not enough categories to recognize all the different levels of relatedness that a modern cladistic "phylogenetic tree" can recognize, and nobody wants extra formal ranks with names like supersubfamilies or subsuperfamilies. There are quite enough names already.

PhyloCode is based entirely on cladistic phylogenies, and provides a system for naming clades—all of them. The old formal Linnaean categories above species level are abandoned. This is a rather revolutionary suggestion, to say the least, and it is not surprising that it has excited some strong opposition. To my mind the strict logic of the PhyloCode is beside the point. The most important thing about the current system of naming organisms is the common language it provides, not just to other systematists, but to the rest of the world—people like gardeners, or bird watchers, or fungus forayers. Very few members of this larger community know about the details of cladistic phylogenetic analysis, and I suspect that most of them want a meaningful label that they understand rather than reassurance that every category is quite the latest collection of good clades. The 250-year tradition since the great Swedish systematist does count for something. Many of the common categories that a naturalist will comfortably recognize are old Linnaean families. Think of lilies (Family Lilaceae) or daisies (Family Asteraceae) or crows (Family Corvidae). These turn out to be pretty good clades as well, meaning that the resemblance between the species in the families does indeed reflect descent from a common ancestor. In my experience more "difficult" groups of organisms are often reanalysed time and again using the latest cladistic bells and whistles or new molecular evidence, and each new analysis is rather different from the last one. Nor is there any guarantee that the latest version is always the best. Potentially all these different analyses could be named under PhyloCode. In my view this would allow for just too many valid names, as each successive analyst sought to put his imprimatur on his briefly dominant hierarchy. But most important of all is a feeling that offends my democratic instincts, in that the systematization of nature would be even more in the hands of a coterie of specialists sitting in front of their computers than it is now. The binomial system has faults, but I suspect any new system would develop as many. The naming process would be taken away from the naturalists, nature lovers and intelligent laymen, at a time when there has never been so much pressure on the survival of species, or, indeed, on the survival of the taxonomists who know about fleas and carabids, trilobites and ammonites, grasses and orchids, or deep-sea worms. It is the survival of the biological world and of the basis of expertise that studies it that is the real concern of the twenty-first century. Names are the least of it.

3

Old Worlds

It might seem an odd ambition to try to get everyone to pronounce a word correctly. But mine has always been to get the world to say "trilobite" without fudging, and with a certain measure of understanding. My own mother was wont to say "troglodyte," which at least has a certain prehistoric dimension, even if it refers to human cave dwellers rather than extinct arthropods several hundred million years older than humans. "Did you have a nice week with the troglodytes, dear?" was one of her regular enquiries. A rather more common mispronunciation is "tribolites"—an anagram of the correct word for sure, but probably an unconscious *hommage* to one of the humanoid tribes on *Star Trek.* "The tribolites have made it through the air lock, Captain. Permission to use phasers!" I have no particular gripe against those who pronounce the word with a first syllable to rhyme with "thrill," although I have always said "try-low-bites" myself. The tri- part, of course, refers to the threefold division into which the calcareous carapaces of these animals are usually obviously divided lengthways—"three lobes." On their underside, but rarely preserved, were many jointed legs of typical creepy-crawly kind, which reveal the trilobites to have been distant cousins of the crabs, butterflies, spiders and millipedes, with which they should be classified—in Linnaean terms, Phylum Arthropoda, Class Trilobita. For getting on for three hundred million years trilobites swarmed in the oceans, moulting and mating, and left behind their hard carapaces in the rocks as testimony to their former importance. At the moment we know something like five thousand genera of trilobites, and new species are being discovered entombed in ancient sediments such as limestones and shales. It is not surprising that they have been described as the "beetles of the Palaeozoic." In fact, they still have a long way to go before they approach the beetles in biodiversity, but they are wonderfully varied creatures despite their simple ground plan, some with carapaces as smooth as beans, others like arthropodan porcupines, many as large as lobsters, yet others as tiny as water fleas. They evolved fast and are not uncommon fossils, so that they are useful in dating rocks—somebody who "knows his bugs" should be able to say within a few minutes whether he is looking at Cambrian or Devonian examples; with more study the time zone can be narrowed further. Trilobites can tell about ancient climates, because different species lived in tropical as opposed to cool seas. They can tell us about vanished continents in distant eras, since different trilobites characterized different parts of the world. Study of apparently esoteric extinct animals can help us reconstruct the history of our planet.

A few years ago I wrote an account of recent discoveries of remarkable trilobites in the Devonian rocks of Morocco, dating from more than four hundred million years ago. I included an illustration of a bizarre creature that carried a trident on its head, as far as I know a unique structure in the whole animal kingdom. In 2000 this trilobite had no scientific name, although it was already possible to buy specimens of it over the internet. My colleague Pierre Morzadec named the trident-bearing genus *Walliserops* in 2001, commemorating a well-known Devonian specialist, Professor Otto Walliser of the University of Göttingen. In 2005 I went to Morocco to see the localities where these trilobites had been discovered. Brian Chatterton of the University of Alberta in Edmonton, Canada, has been making a study of the trilobite sites in the Anti-Atlas for several years, and he invited me to join his field party that spring. Trilobites have become a major industry in the area around the small town of Erfoud, which had previously subsisted on a little bit of tourism to add to the small rewards provided by dates and agriculture. Mr. Hammi was our guide and mentor; he is a helpful local Berber with no scientific training but a hugely intelligent "eye" for a good trilobite. Since these trilobites have to be laboriously prepared out of the rock by hand, Hammi has made a successful family business with his brother supplying splendid specimens that have finished up in collections around the world. He has partly been equipped with microscopes and tools by an English fossil dealer from Cambridge, Brian Eberhardie, who also sold on some of the best examples. I might add that there are several other successful businesses producing cheap fakes. The localities in question are dotted over one of the most barren desert regions in which I have ever worked. It seems that the High Atlas Mountains steal all the water, leaving but a trickle for the Sahara side. But the barren hillsides of the Anti-Atlas provide perfect exposure of the rock formations. This is what geologists call "layer cake" stratigraphy, where each stratum is horizontal, or gently tilted, so that climbing up a hillside from stratum to stratum is equivalent to climbing a staircase upwards through geological time. A productive layer can be traced over a long distance, or at least until a stretch of stony desert interrupts the outcrop. We went to an isolated hill called Zguilma, where trident trilobites had been collected over several years. There was actually a tree or two at the foot of the hill, tapped deep into some tiny source of water. Even in the cooler part of the year their shade was difficult to resist at three in the afternoon; in the summer it must have been impossible to work in the open.

The extraordinary sight that greeted us at Zguilma was a trilobite mine. The productive layer had been traced all along the bottom of the hillside and dug out in a series of trenches and pits, flanked by piles of debris. When Hammi arrived, muffled shouts in Arabic sounded from a hole, and out climbed a cadaverous old man with one or two yellow teeth displayed in a broad grin. He had been ten feet down in the hole in the full heat for several hours breaking hard limestone rocks. It was like being employed in Hades, with added hard labour. Mysteriously, the old man seemed cheerful enough. He was the beginning of the chain of discovery, for if he broke across one of the precious trilobites he would put both "halves" on one side, and Hammi would pay him modestly for the find. Then it would be taken on to the laboratory for preparation, and if

Trilobite "mines" in the Devonian strata of the desert in the Moroccan Anti-Atlas

a good trilobite were extracted, might eventually fetch up at the Houston Fossil Show or some similar event carrying a price tag of several thousand dollars. The wizened old man seemed untroubled by the chain that led to Houston, and was doubtless unaware of the profit differential; he was glad of a break to share the sweet mint tea that is the social lubricant in the desert ("Berber Whisky" is a joke for the infidel). Every evening, in the incomparably still dusk that comes in the desert, we would all share *tajine* made from tough old bits of meat that had spent the day hanging on string between the branches of the token tree, mopped up with Moroccan bread cooked in warm embers. We had the same bread for breakfast spread with "La vache qui rit" processed cheese. After a couple of weeks the diet began to pall. I have been allergic to laughing cows ever since.

Several years earlier I had persuaded the Natural History Museum to purchase from Brian Eberhardie another extraordinary trilobite from Zguilma. Now I had the chance to examine where it had come from for myself: evidently, it had emerged from some hellhole. The eyes on this animal were like those of no other trilobite, because they were elevated into a pair of near vertical towers, the outer side of which were lined with very conspicuous files of lenses. Sight was obviously at a premium for this particular species. The challenge was to work out how such flamboyant "peepers" worked. One thing could not be disputed: this heavily armoured trilobite bearing its massive eyes must have lived on the sea floor. I then noticed something curious about the eyes: they had eyeshades overhanging them. Most trilobite eyes are rather strongly curved from top to bottom, with numerous tiny lenses and no eyeshade, but this trilobite had relatively few lenses in a vertical array, so that they looked like ranks of windows in the tower. This is an appropriate simile because trilobite lenses do indeed work as a kind of window made of the mineral calcite. Because of the optical properties of this particular mineral, light passes through the lenses normal to their surface, or, to put it another way, it is possible to tell in which direction a given lens could see by imagining a ray of light impinging at right angles to its surface. So it was obvious that this remarkable trilobite could look all around over the sediment surface on which it dwelt, for the lenses were arced in a semicircle in each eye affording a "view" of the surrounding area. It obviously could not look upwards, not least because the eyeshade would inhibit the view in that direction. Then again the vertical arrangement of the lenses meant that the trilobite could see distant objects. The curved nature of most trilobite eyes means that each lens subtends a cone of sensitivity that naturally widens the farther away from the eye you are; the sight was good close by and poorer at distance. By contrast, the big-eyed trilobite with its straight-sided eye would have been able to detect small movements in

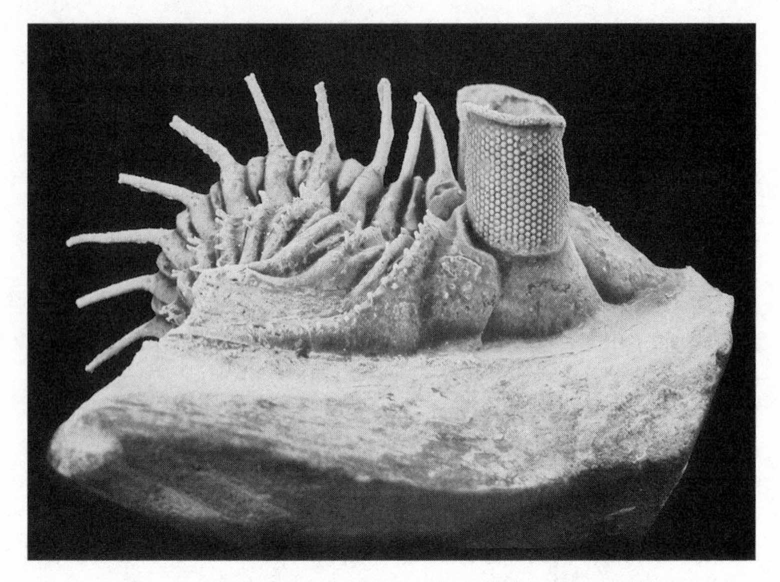

The remarkable eyes of the Devonian trilobite *Erbenochile,*
seen from the side

prey even at some distance. But there is a problem here, for distant light is also weaker, and interference from stray rays becomes more of a problem. This is where the eyeshade comes in. For it rather neatly cuts out the light from above which affords the greatest distraction for shallow marine organisms (at moderate water depth light is refracted to come vertically from above). It is rather like a hunter on the African plains contemplating a distant impala by shading his eyes. The trilobite anticipated the baseball cap by four hundred million years.

This was an exciting enough discovery to publish a short account of it in the journal *Science.* The first thought I had was that this *must* be a new kind of trilobite. My mind began to ferment all sorts of nice descriptive names—*Gogglyops* or *Spectaculaspis* perhaps. I had already checked through the trilobite collections to see if there was any specimen that resembled ours—and there wasn't. But before I got too embroiled in new names I discovered an extremely obscure publication about some rocks in Algeria. Not many libraries have copies of the *Notes et Mémoires,* but the Natural History Museum is one of them. It was

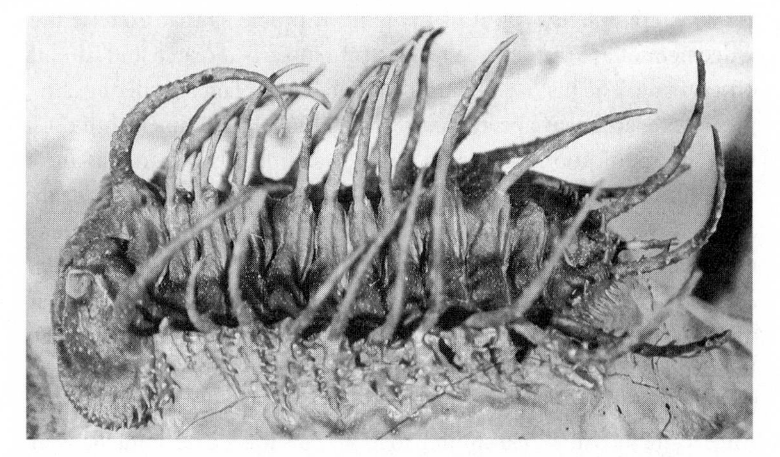

Another astonishingly spiny fossil trilobite from the Devonian of Morocco: the spines on this odontopleurid are genuine, but fakes are often offered for sale.

clear from a rather poor illustration published in this journal in 1969 that a trilobite similar to ours had been collected across the border in Algeria, not very far as the desert crow flies from Zguilma. Clearly, we needed to know more. Fortunately, we discovered that Pierre Morzadec had refigured this material in a rather less obscure journal, and he had dug it out, or as we say prepared it, rather well from the rock, so that one could see more of its features. He had also given it a new generic name, *Erbenochile.* However, all the material from Algeria lacked the head, surely the most distinctive part of the trilobite. But close examination revealed that the tail of the trilobite was almost as distinctive as the head, having a very particular pattern of spines around its margin, which was different from that of any other Devonian trilobite. The Algerian specimens were identical to the Moroccan one as far as the features of the tail were concerned. There really was no escaping the fact that our spectacular trilobite species had been named already, albeit from a specimen lacking the remarkable eyes. Applying the rule of priority means that *Erbenochile erbeni* is the name we must use for our trilobite. If we had not had access to a wonderful library, we could well have got the name wrong, and caused much confusion for future generations.

This example is typical of the kind of problems that exercise the judgement of a taxonomist, a mixture of scholarly research and careful observation. The history of naming animals and plants is full of examples where labels have been incorrectly applied. In the nineteenth century communication between scholars was imperfect, so it was then quite likely that an animal or plant might have been named twice by accident. The priority rule often had to be applied. I regret to say that there were also numerous cases where scientists "rushed to press" to establish their priority over any potential rivals. One of the most infamous examples concerning fossils was the race between Professors Edward Drinker Cope and Othniel Charles Marsh in the latter half of the nineteenth century to describe and name the spectacular North American dinosaurs then coming to light. This was a case of intellectual war, fought out in publications and in academic disputes. The two protagonists really loathed one another, and each was determined to name any newly discovered animal before his rival. Such enmity certainly stirred up a fever of activity in the prosecution of the war of reputations, but sometimes the casualties were names that got caught in the crossfire. In other examples, it is hard to establish who or what has priority, and the bemused scholar will find himself examining the small print on inside covers to find out whether a given book was published in May or September of 1799. I have used faded library stamps as evidence of the receipt of a publication by the Museum—which must therefore have been published earlier in its country of origin. What is evidently needed is a set of laws to sort out nomenclatural disputes—and so we have the International Code for Zoological Nomenclature, and there is a botanical equivalent. I have to admit that the Code makes for pretty dull reading and can, in the wrong hands, become a pedant's playground. But it generally works to sort out which name is the valid one. However, there are cases when a rigid application of the Code would result in something silly happening to very familiar names. This might occur, for example, if some bookish scholar discovered a work of unprecedented obscurity containing earlier names for well-known animals. It would be highly undesirable in this case rigidly to apply the rule of priority, for names are a means of communication first and foremost, and nobody wants to revive an old name just for the sake of it.

But how can a zoologist decide when to flout the rule of priority? The answer is to apply to the International Commission on Zoological Nomenclature (ICZN) with details of the case in question. With sufficiently good reasons a later name might well be conserved—this is decided by a vote of the Commissioners, who are an international group of taxonomists. Mostly this is just a way of formalizing common sense.

But it is necessary to have a court of appeal on names, because there are millions of species of animals still to describe, and it would be amazing if there were not more nomenclature conundrums thrown up, quite apart from having to continue to deal with problems arising from the last 250 years of scientifically naming animals.*

Lack of glamour was never a problem for *Archaeopteryx.* This is probably the single most famous fossil in the Palaeontology Department: "the first bird" is how it has been referred to for generations. Its full scientific name is *Archaeopteryx lithographica.* The species name refers to its occurrence in the Jurassic lithographic limestone mined around the towns of Solnhofen and Eichstatt in Germany. It is exposed in a very large open quarry where thin, flat slabs of nearly white limestone can be prised up from the outcrop as if they were stacks of tiles. As its name implies, this fine-grained stone is perfect for making lithographs. Occasionally, the quarry yields the laid-out skeleton of a fossil on a slab, as perfect as a picture at an exhibition. Most commonly these fossils are fish, but very, very rarely an *Archaeopteryx* will turn up. The eight specimens known so far represent the yield from millions of tonnes of rock. The rock was originally laid down in a lagoon, but one that was fatal to most animals because of its low oxygen content—which also protected potential fossils from destruction by scavengers. Fish that swam in from the open ocean were both doomed and immor-

*After the Convention on Biodiversity following the World Summit in Rio de Janeiro in 1992, the signatories agreed to move towards conserving their biodiversity at a national level. This means knowing what animals and plants you have got, and recognizing how to name them properly. Standardization of nomenclature is part of the process, right at the start. The problem is one of investing nomenclature with glamour, and I have to say that it does have a whiff of accountancy about it. *The Journal of Zoological Nomenclature* could never be described as a riveting read, even by its devotees.

talized. Where sticky lime banks rose above the water level creatures like *Archaeopteryx* might get trapped—there are also several species of flying reptiles, or pterosaurs, from the same deposit. The critical specimen at the Natural History Museum was discovered in 1861, and Richard Owen secured it for the then British Museum for £700. That was a very large sum of money for the time; it is misleading to make simple conversions, but consider that a house could be bought then for about a hundred pounds. Owen had realized quite quickly the importance of the new fossil, and gave it its name and first description in 1863. The type specimen, the holotype of Owen, is for ever BMNH 37001. The bantam-sized fossil is spread-eagled on its slab like roadkill, feathers splayed, a buff skeleton on a cream background. It is kept out of the way within the department in secure storage: the specimen on display in the bird gallery is a carefully made replica. I think we could probably scale up the value of the original today in proportion to Owen's purchase price, so a measure of caution is prudent.

You must not get the idea that once the specimen has been labelled "holotype" and filed away in its appropriate box behind the scenes that means the end of its active life, apart from an occasional dusting. Nothing could be further from the truth. The specimen is available for re-examination and reinterpretation. No scientist ever has the last word, much as he might like to think he has. Historical reality is continuously remade, as ideas evolve and change. The geological past is a construction of the mind: study of specimens is worthwhile if it serves to push interpretation a little further in a new direction. No doubt when Owen described *Archaeopteryx* he thought he had made as much of it as he could, and he had certainly realized that it showed the combination of bird and reptilian features that has given it such an important place in the history of palaeontology. But as other specimens were discovered, further details were noticed. Refinements in preparation techniques allowed yet other features to be excavated; the late Peter Whybrow spent many hours using a mechanical pin under a microscope teasing out buried bits of bone on the type specimen. A dozen experts of their day have given the same specimen of *Archaeopteryx* a close inspection, and you might be forgiven for thinking that anything that could be extracted from it must have been by now. But the importance of the type specimen has been revived yet again in the last few years since the discovery in China of feathered dinosaurs. Feathers, it transpires, are not a uniquely avian characteristic after all. While this adds greatly to a theory that birds are descended from dinosaurs, it does open up the possibility that *Archaeopteryx* is not, after all, the first bird of common knowledge, but actually another slender dinosaur, a coelurosaur, which was on its way to becoming a "proper" bird. Or, which is the same question in different dress, what are the characteristics of a bird, if not feathers alone?

Modern anatomists recognize very well that there are a number of features that *Archaeopteryx* shares with relatively small and graceful coelurosaur dinosaurs; these are mostly primitive characters. While this may tell us about the ultimate line of descent of the Solnhofen flyer from some kind of diminutive inhabitant of Jurassic Park, what we are really looking for are advanced characters that *Archaeopteryx* shares with true birds, particularly if we wish to know whether birds + *Archaeopteryx* constitute a natural clade, such as I described in the last chapter. Teeth, for example, which *Archaeopteryx* has in abundance, are neither here nor there as far as its classification is concerned, because they are merely a feature retained from its dinosaur ancestor. Much more significant is the fact that the asymmetrical flight feathers on its wings and tail resemble those of modern birds, and they are not like the feathers of feathered dinosaurs. To use a hackneyed phrase, they are "fit for purpose" for flight. What was needed now was new information to clinch the matter, and that meant going back to the specimens yet again. Out came the slab bearing Owen's holotype once more. This time, Angela Milner and her collaborators were trying out a new technique on the old specimen. They were looking into the braincase of the feathered fossil using X-ray CT (computed tomography) technology—well, if it can peer into the intimate spaces of human anatomy, it ought to work on an old bird. The results were beyond all expectation, and sufficiently startling to earn a place in *Nature* in 2004. The palaeontologists were able to model even fine details of the brain, demonstrating beyond question that the Jurassic animal had a highly developed visual system and expanded auditory and spatial sensory perception. They even managed to produce a model of the inner ear of *Archaeopteryx*, an object only a couple of centimetres long. All the evidence was consistent with *Archaeopteryx* as an animal capable of powered flight—not perhaps the equal of the aerodynamic marvels of the modern era, but nonetheless a fully effective flyer. If it looks like a bird and flies like a bird, then, by golly, it *is* a bird. Why, it even thinks like a bird!

So now the famous specimen is back in its drawer once again. Its records will have been updated with the latest details of scientific publication—all rather dry notes compared with the real excitement of the new discoveries. We know a little more about the intimate daily life of what was almost certainly a real bird. We also know that feathers had been acquired by dinosaurs before their avian descendants ever took to the air. Feathers are evidently another example from the fossil record of what Stephen Jay Gould called exaptation—a structure appearing for one purpose and then being recruited for another. Feathers may originally have been selected for insulation. It goes almost without saying that not every researcher will agree with what I have just said. There is still a small but vocal school that disputes the relationship between dinosaurs and modern birds. Science is all about disagreement. It is likely that the "BM *Archaeopteryx*" has not made its last public appearance. After all, nobody can anticipate what new techniques may yet be applied to the unfortunate bird that found itself trapped in calcareous mud in an ancient lagoon sweltering under a hot sun in what would, one day, become Germany.

The examples of *Erbenochile* and *Archaeopteryx* will demonstrate that discoveries officially become recognized only when they are published. This is true of science in general, and natural history provides no exception. Up until the point of publication a discovery is only provisional, and most scientists enjoy a certain measure of nervousness while a scientific paper is in that limbo known as "in press." Some unscrupulous rascal might yet sneak into print ahead of you! Scientific names are the least of it. Nowadays, it is publish or perish, *Nature* red in tooth and claw, unnatural selection—and, to the winner, the spoils. The description of a scientist, no matter how brilliant, as a "non-producer" is a very effective way of blocking his or her promotion. The result is that too many papers are published, and many scientists are so busy keeping ahead of their rivals that they don't even have time to read what other workers write.

Life as a scholar was not always so fraught. When Leslie Bairstow was appointed to the Department of Palaeontology in 1932, the Keeper of the day was reported as rushing through the offices crying excitedly: "*We've got Bairstow! We've got Bairstow!*" He came from Cambridge trailing glory after a brilliant undergraduate career, having been appointed to a Fellowship at King's College at a prodigiously young age. He moved into a safely tenured position at the BM and proceeded until retirement to publish—nothing. Not that he was idle. He had started researching assiduously the strata of the Jurassic Lias Formation around Robin Hood's Bay in Yorkshire. The shales and impure limestones of the Lias are beautifully exposed in cliffs and along the foreshore in the bay close to the charming old fishing town of Whitby. The commonest fossils from the Lias are probably the coiled molluscs known as ammonites, which are also used for dating its successive levels. Handsome, glossy specimens of the genus *Dactylioceras* are commonly sold in rock shops, and not just around Whitby. Local legend has it that they were originally snakes turned to stone by the Abbess of Whitby, St. Hilda, and some were even embellished with carved snakes' heads to add authenticity to the story. By now it will be no surprise to learn that St. Hilda, too, has a genus named for her—*Hildoceras* (ammonites conventionally have the suffix *-oceras*). Bairstow devoted his attention to another group of fossils, belemnites, which can be collected throughout the Lias. They had been rather neglected, and perhaps this is not altogether surprising, because they are, frankly, rather dull-looking fossils. Their appearance is often compared to that of cigars, and indeed they do range in size and shape from small cheroots to generous Havanas. These common fossils are just the calcite "guards" of squid-like molluscs, and remains of the whole animal are hardly ever found. So the palaeontologist is working with scraps. Bairstow collected belemnites from layer upon layer of the Lias, and meticulously recorded their occurrence. This did not satisfy him, and so he had to return to collect on an ever finer scale, thus acquiring many more specimens, all of which needed curation and filing. He developed a reference system that operated with cards and knitting needles, a kind of Heath Robinson precursor of the punch-card filing system. His collections grew and grew. It is said that he filed everything. When he was sent specimens in parcels to identify, which he did with great thoroughness, he would unpick and save all the bits of string with which they had been trussed, and file them all according to length in special boxes. When he retired there were discovered a number of such boxes, labelled "string: 2–3 feet" and so on. One box was labelled "pieces of string too small to be of use."

Eventually, his room was so crammed with specimens and paraphernalia that it was arranged as a kind of maze, through which the visitor had to pick his way—and his approach could be heard well in advance. At the centre of the maze, like a spider in its web, Bairstow sat wearing green eyeshades, playing with his knitting needles or filing his bits of string. He pretended to be very busy by the time the visitor arrived. His lack of publication was attributed to a paralysing perfectionism. Even collecting the Lias centimetre by centimetre would never suffice; nothing was ever quite in the condition necessary for writing up. In fact, he did once publish something—by accident. One of the replies he had written to an enquiry appeared in print in *Happy Days*, an amateur cycling journal too obscure even to find its way into the library of the Natural History Museum, although I have seen a faded copy of the article, which deals with a fossil that was brought in as an enquiry: I am afraid it is of no consequence. By the time I joined the Museum, Bairstow had retired, but still appeared at the end of the working day like a sad wraith—tall, thin, moustachioed and impeccably polite—just as everyone else was going home. He had taken up deep-sea diving, so that he could gain access to the submerged strata of his beloved Lias on the Yorkshire coast. When he adopted this occupation, diving suits were very heavy, with weighted boots and a bolted-on helmet with a glass visor. He first tried on the suit after hours at the Museum, found himself unable to get out of it again, and was forced to leave the Museum to get help. The stricken scientist was compelled to plod up Knightsbridge, like an extra in a science-fiction film, mouthing through his mask at strangers until he found somebody to release him. The story of Bairstow might seem one of promise unfulfilled, but it has a happy ending of sorts. After his death, his successor at the Natural History Museum, Dr. Michael Howarth, wrote up the Jurassic stratigraphy of Robin Hood's Bay from Bairstow's notebooks. Finally, he had broken into print.

Stratigraphy was a popular field of study in the twentieth century; the word means "the drawing of strata"—and many of the scientists working in the department then would have thought of themselves as geologists and stratigraphers first, palaeontologists second; indeed, this part of the BM changed its name from Department of Geology to Department of Palaeontology only in 1956. Bairstow's unpublished work actually followed a tradition of research that went back to the earliest days. Possibly the most historically important stratigraphical collection of fossils that exists anywhere was purchased between 1816 and 1818. This was the collection of William Smith, who produced the first accurate geological map of the strata of England and Wales—and was among those who laid the foundations of stratigraphy as a science. The splendid map itself is on display on the walls of the Geological Society of London, the oldest society of its kind anywhere, although that organization only belatedly recognized Smith's great contribution: after all, he was "trade" rather than a gentleman. The tradesman won his proper measure of fame in the end, and Simon Winchester has done his best to add "Strata Smith" to the list of nineteenth-century scientific heroes in his book *The Map That Changed the World.* Smith had used the succession of "organic remains" recovered from the strata he crossed when he was a canal surveyor as a fingerprint for the geological formations. Particularly important were ammonites (see p. 89)—they were easy to collect, and their various patterns of whorls and ornament were as recognizable as the faces of old friends. The specific name of *Clydoniceras discus* tells you all you need to know about the appearance of this particular species, and it will be found only in the lower part of the thin limestone formation known as the Cornbrash, which William Smith had been able to map when he traced the Jurassic rocks across country. Follow the stratum and find the fossil. His collection of about two thousand specimens was a conflation of teaching aid and practical guide. It is kept together now in one cabinet, and some of the writing on the labels must be that of Smith himself. Many of the ammonites are fine examples by any reckoning, and they give meaning to that old description of a sample as a "hand specimen"—for they do indeed fit comfortably in the hand, like a medal, or a sports ball. The Jurassic sea urchin *Clypeus ploti** is more like a well-baked bun. It is an extraordinary thought that William Smith would have shown one particular black *Dactylioceras* ammonite to a convert to his stratigraphic method. See how it has acquired a patina from handling. Teaching by example leaves a subtle shine.

Perhaps Bairstow was inhibited in publishing by the presence in the same department of L. F. Spath, who did little else. He was a kind of ammonite-describing machine. He published on ammonites from around the world: monographs on the ammonites of the Cutch appeared in the *Palaeontographica Indica;* he published on ammonites from Folkestone, from Argentina, from Skye. His output was vast. It must have been intimidating to have such a prodigious character working in the same department. One of my senior colleagues actually knew Spath in his latter years. He reports that Spath had a memory that was so encyclopaedic he could cite the very page on which some particular ammonite had been figured, no matter how obscure the journal. In the end he became quite blind, and he gently stroked the specimens to feel the outlines of their diagnostic ribs and tubercles, and could identify them from touch alone. Oddly enough, he never had a fully "established" job at the Natural History Museum, and continued to be employed on a part-time basis even after he had a worldwide reputation. This may have been because he was an autodidact, having taken his several degrees as an external student. It is unlikely that it had to do with his German origins, because he had served with the Middlesex Regiment in the First World War, but it is said that the Museum was so stingy in its job offer afterwards that he preferred to supplement his livelihood elsewhere. This included teaching at Birkbeck College as a temporary lecturer. Birkbeck is a wonderful institution for devoted students who work for degrees in the evenings, usually after a hard day's work elsewhere. Spath also did what was effectively "piece work" on

*This species is named for Robert Plot (1640–96), who was a pioneer palaeontologist and the first Keeper of the Ashmolean Museum in Oxford. His *Natural History of Oxfordshire* figured a number of fossils of Jurassic age for the first time, including the bone of a dinosaur, later to be named as *Megalosaurus* ("huge lizard"). Clypeus is a Roman round shield.

Ammonites like this Jurassic dactylioceratid from Lincolnshire are not only beautiful but useful in the correlation of sedimentary rocks—the science of stratigraphy, which dominated life in "Palaeo."

ammonites. The journal *Palaeontographica Indica* was established to publish on the fossils of the British Empire in the east and especially India, so he was paid almost by the page, or by the species, to record discoveries from the Himalaya and elsewhere in the subcontinent. There is a view that this encouraged him to recognize rather more species than was strictly necessary.

Nonetheless, you cannot argue about Spath's sheer industry. He was elected to the Royal Society in 1940, and must have been regarded as near the top of his particular tree in this, the heyday of ammonite studies. His only rival was W. J. Arkell of Oxford, author of the monumental *Jurassic System of Great Britain* (1933) and the even more monumental *Jurassic System of the World* (1956). Not surprisingly, the two authorities did not like one another. But, thanks to them, ammonites became the supreme fossil chronometers used to subdivide Mesozoic geological time, the stratigrapher's ideal material. Almost every few feet of strata could be typified by the appearance of another species of ammonite. Because of their rapid evolution and widespread geographical occurrences, ammonites were a boon to any geologist in the field almost anywhere in the world for determining the age of the strata before him. Even in England, thanks to ammonite precision dating, it was possible to detect earth movements that had hitherto been obscure. It is probably also true that ammonites have been slowly declining as objects of study ever since the time of Spath, as other fossils, particularly microfossils, have taken over their role as chronometers. There are fashions in fossils, as in every other aspect of our culture. In Britain today the only full-time paid professional expert on ammonites is the rubicund Professor W. J. Kennedy of Oxford University Museum, another sterling labourer at the monographic coalface, and one who has the persistence to carry on working even though he may be the last of his kind, like the pearly *Nautilus*—last living genus of the Order Nautiloidea which included the ancestors of the ammonites themselves.

Spath might serve as the model for many of his contemporaries in the Department of Geology. Great cataloguers and stratigraphers, they worked on corals or brachiopods or echinoderms or clams, and became old-fashioned authorities. Several of them laboured over collections studied by still greater forebears. T. H. Withers, for example, prepared a *Catalogue of Fossil Cirripedes,* which included many specimens studied by Charles Darwin himself. Cirripedes are known to most of us as barnacles—anomalous crustaceans that have taken a liking to encrusting rock surfaces, or whales, or one another. They feed by means of modified limbs which protrude from the calcareous valves that enclose the body, like so many feathery nets spread to catch their microscopic food. Barnacles are found from the upper tidal zone to the deepest ocean, so there are plenty of species. There are two main kinds: encrusting sessile ones—acorn barnacles—that are familiar for making some seaside rock surfaces excruciating to walk over, and the goose barnacles with flexible "stalks." Let us open one of the drawers containing Darwin's specimens. A series of blocks lie neatly arranged in trays. Most of them are a little disappointing because only the calcareous plates that comprise the casing of the cirripedes survive, and then they fall into individual plates on the death of the animal. A grey piece of chalk has a few of the individual valves scattered over the surface, in shape like miniature kites, but it should be possible to piece together the original barnacle if we know enough about the arrangement of plates on living species. That sloping writing on the label is probably that of the great man himself. Some naturalists will be unaware that Darwin spent eight years during the 1840s devoted to the study of barnacles, both living and fossil. He published a monograph on the fossils in one of Britain's oldest scientific journals, the *Monographs of the Palaeontographical Society.* He used the time to mull over and develop the thoughts that would make him famous when *The Origin of Species* was published a few years later. Rebecca Stott has shown that the barnacle years were important to establishing Darwin's respectability as a serious zoological researcher as he made the transition from his earlier phase as a geologist. The scientific basis of the barnacle classification he developed was sound enough to be employed even today in its essentials. Like all taxonomists, he had noticed a mess and set about sorting it out. As the principles of classification clarified as he studied more specimens, so, too, did the arrangement of the ideas that would come to fruition in evolutionary theory. The acorn barnacles helped the growth of the many-branched oak tree of modern biology.

The Museum used to be remarkably hierarchical. The scientists talked to one another and only rarely hobnobbed with their assistants, who were known as Experimental Officers. There are a few survivors from this era. As this is written, Ellis Owen is eighty-three and still working on the shelly fossils of brachiopods, an occupation that seems to have preserved him from the normal processes of decay. He worked for Helen Muir-Wood, doyenne of brachiopods, who was known universally as "Auntie" (though not to her face). A redoubtable maiden lady, she died in 1968, having risen through the hierarchy to become a Deputy Keeper, a trailblazer for her gender. Ellis recalls being treated rather like a slave. He says that the hierarchy was extremely rigid, just like officers and men in the army. The scientists had their own common room, into which the Experimental Officers dared not venture. Meanwhile, the army really did supply the warders on the public galleries until the 1970s. Ex-army personnel were considered just the thing to keep the visitors in order, and stiff, blue, military-looking suits were the right uniform in which to do it. The warders were ruled over by a hard-drinking Scottish Head Warder known as Mitch ("Mister Mitchell to you"), who loudly briefed his subordinates en masse every morning in the Main Hall, shouting at them just like any regimental sergeant major. Most of the warders scowled a lot of the time. Now they are trained in people skills, smile routinely and have nice designer uniforms. It's hard to feel nostalgia for the old days.

When I joined the Natural History Museum in 1970, the old order still persisted in a somewhat more benign form. I was admitted to the Senior Common Room in the basement corridor even though, as the dinosaur man Alan Charig put it, I was "still wet behind the ears." An agreeable fug of cigar smoke permeated the room, which boasted a coffee maker and some passably comfortable chairs, occupied by perhaps a dozen scientists. By then, the sports jacket was the uniform, but for a while I attempted to follow the fashion of the day, with a black, wet-look jacket and a red shirt with a big white collar. In the end I, too, adopted the tweedy option; trendiness withers in the vaults. Anecdotes drifted around the Common Room along with the plumes of smoke, but there was not much talking shop, as if it were considered bad form to wax too enthusiastic about the latest discovery. There was still an unspoken sense of hierarchy, and everyone in the room knew exactly which rung of the ladder you were on. There were, of course, no Experimental Officers in sight. The scientific grades went from Assistant Scientific Officer, to Scientific Officer, to Higher Scientific Officer, Senior Scientific Officer, Principal Scientific Officer, Senior Principal Scientific Officer and, somewhere in the stratosphere among the cirrus clouds, Chief Scientific Officer, or the Director. These posts were all known by their acronyms—SSO, PSO and so on—you knew that nobody under SSO would be allowed in the Senior Common Room. Promotion was a long slog upwards through a welter of acronyms. I give you this tedious litany of titles to show how structured the scientific Civil Service used to be. I heard a story from one of the laboratory staff, Johnny Meade, about life under the Keeper of Geology, W. D. Lang, who "reigned" from 1928 to 1938. Meade had joined as a very young assistant, and when I met him was near the end of a long career. Lang would require that an Experimental Officer take out his microscope every morning and place it upon his work desk. On one occasion the usual assistant was ill, and Meade was dispatched belatedly to perform the morning ritual. "You're late!" sniffed the Keeper. When Meade explained the circumstances and offered to bring out the microscope to its usual place, Lang regarded him loftily. "It is too late. I shall not be requiring my microscope today . . ." Such was the iron routine in the days of rigid hierarchy.

I have mentioned that nearly all the contemporaries of Muir-Wood, Spath and Withers in the Department of Geology were greatly interested in stratigraphy—in using fossils as tools for correlating rock successions. This interest extended to reading the rocks for direct evidence of evolution. Put the fossils in order and the course of evolution would be revealed like a petrified narrative, and the ticking of geological time could be measured by the same token. If some unexpected fossil appeared at the wrong level in the rocks—well, that meant that not enough collecting had been carried out. Then it was back into the field doing much damage with the geological hammer until all became clear. Spath collected the ammonites from the Cretaceous Gault Clay metre by metre in a manner almost Bairstow-like in its thoroughness, and spent several decades publishing the results. The evolutionary narratives that were allegedly revealed by the stratigraphic studies became the basis of a variety of ungainly scientific terms—palingenesis, lipo-palingenesis and the like—that have now been almost entirely forgotten. Palingenesis, for example, described the appearance in an immature descendant species of features that were found in *mature* individuals of an ancestral species in strata immediately below it. Many modern workers believe that such graded, simple evolutionary sequences from one species to another are a rare occurrence in the rocks. Species remain comparatively unchanged through a thickness of strata and then are replaced rather suddenly by another. However, there has been a revival in interest in the kind of changes with which the ammonite specialists were concerned—but by now this work has developed an entirely new vocabulary! Regardless of the debates about evolutionary theory, the scientific names of the ammonites live on, as do the subdivisions of geological time that they helped to distinguish. It is sometimes difficult to anticipate which contributions will stand the test of time. This is a telling example of how facts tend to endure whereas the ideas that the facts originally engendered are more subject to revision or reinterpretation.

The Experimental Officers were released from their slavery when a reform in the Civil Service abolished the difference between them and the Scientific Officers in the early 1970s. Hard-working people like Ellis Owen could now rise into the ranks of the scientists on the basis of merit—and they did just that. The same period marked a rejection of the tradition of the stratigraphic palaeontologists that had held sway almost since the fossils had their own department: these were revolutionary times indeed. It was surely a good time to throw over the shibboleths of the past. I found myself in a kind of limbo, since I had been raised in the tradition of stratigraphy, and now found myself required to examine my principles. In the previous chapter I briefly introduced the cladistic methods that were to take command of the taxonomic agenda. The high priest of cladistics was a palaeontologist in our department, Colin Patterson, who worked on fossil fishes. He was one of the few scientists I have known gifted with that mysterious property called charisma. Many scientists of real distinction would pass unnoticed in a crowd or be the last person to be served at the bar. Colin had one of those voices that instantly commanded attention: a trained Shakespearean actor's kind of voice, not exactly fruity, but with a natural authority such as a good actor might employ to portray Agamemnon or Henry V. He had a brain to match the voice. The odd paradox was that despite his charisma and expertise on fossil fishes, Colin Patterson was dead set against the notion of "the authority"—he insisted that the evidence for natural classification should be objectively based in morphology, a list of features incorporated into a cladistic analysis, which must be laid out for all to see, not buried in the mystique of the sage who knows all, speaking from his Olympian redoubt. I could not help but remember stories of Spath, who was known simply as the Great Man, and whose word was not capable of being challenged—except, of course, by Dr. Arkell. What Colin helped develop was a language—one that is still in use, and will remain in use, unlike the terms coined by Spath and his colleagues.

That language is now the familiar argot in the recognition of clades: for example, a *synapomorphy* is a character shared by taxa that helps to define a group, like the particular feather structure or brain of *Archaeopteryx* I have described that serve to link it with the birds. Its opposite is a *plesiomorphic* character—one that is retained from a com-

The renowned fish specialist and classification theorist Colin Patterson, relaxing in ornithological mode

mon ancestor, like *Archaeopteryx*'s teeth, or the egg-laying habit of the birds as a whole, or the genes that instruct the sequence of our own embryonic development much as they do in the fruit fly. I have shown that you cannot define birds as animals that lay eggs, any more than you can claim *Archaeopteryx* for the reptiles because it has teeth. The idea that classification should be based on synapomorphies rather than plesiomorphic characters whenever possible informs modern systematics, and Colin Patterson was one of those who wrought this clarification. He had started out in the long-established tradition of the Museum, publishing a huge monograph on the fishes of the Cretaceous Chalk. He was always a great believer in the taxonomic purpose of natural history museums, and in the primacy of collections. He also, whether he liked it or not, became an authority. When he delivered a lecture, young scientific visitors used to cluster at the door to hear that commanding voice lay out ideas with exceptional clarity. He pioneered a style of shabby chic that was much copied by the younger generation, in which a sagging velveteen jacket played an important part. The charisma was palpable.

Almost as impressive was Dick Jefferies, who had joined the Museum as an echinoderm expert in the 1960s. Echinoderms are a great group of marine animals that includes sea urchins, starfish and sea lilies. Dick spent much of his subsequent career attempting to remove some peculiar fossil animals known as carpoids from the echinoderms to a position at the base of the group of animals to which you and I belong—the chordates, of which the familiar vertebrates are a part. Dick had the look of the perfect scientific intellectual, with a high, bald cranium and a magisterially distrait manner. His deep sonorous voice is still as distinctive in its own way as the Pattersonian tones, and he has splendidly crested eyebrows that somehow help to invest his remarks with seriousness and just a touch of quizzical humour. The hegemony of the stratigraphic palaeontologists was already a memory, although the Keeper at the time, Dr. H. W. "Bill" Ball, did hail from that tradition. My near namesake, Peter Forey, coelacanth expert supreme, joined the Museum shortly after I did, and we have been receiving one another's mail ever since. The combination of Patterson, Jefferies and Forey was a formidable one in the cause of cladistics. They lived close together on the first floor of the "new building"; their convictions made them slightly scary to other mortals. They tended to refer to me and some of my colleagues as "the stratigraphers on the third floor." If an opinion was offered that did not emanate from the cutting edge, they were wont to dismiss it with "We don't do it like that any more," which simultaneously had the effect of excluding the interlocutor from "we" and making him feel hopelessly out of date. Colin Patterson managed a kind of dismissive sniff, often without looking up from his microscope, which reminded me of the story of Diogenes greeting a visiting dignitary with the cry "Get out of my light." Of course, they were right: we really don't do it like that any more. It was only a few years before desktop computers were constructing trees in a few seconds that would have taken weeks to work out by hand. Systematics had become a modern science.

But does that invalidate expertise? After all, any system that strives towards objectivity should be manageable by technicians and computers, like the sequencing laboratories that routinely examine DNA "fingerprints." Surely, we don't need specialists any more . . .

Actually we do, because the familiar adage "rubbish in, rubbish out" applies even more in an age of systematics married to computer algorithms. The most important part of understanding any group of animals and plants remains their characters, their morphology. Discovery of new, unsuspected characters, new kinds of "hairs on legs," can change a classification for the better, something which remains more important than the latest whiz-bang computer program. The worry now is that the leisurely ways of the old museums are no longer possible. The pressure on publication means that young scientists cannot spend a decade learning their trade—their group of ammonites or moths or whatever; and there is really no substitute for experience. It was always true that some people have a kind of feel for what is significant in a group of organisms; this is a gift, like a musical ear. This gift is found in the born naturalist, and may determine who eventually earns a living in the business. It is equally true that there is the equivalent of tone deafness, which I might term taxonomic blindness. Many people will wave vaguely at a landscape admiring the trees, but won't distinguish an ash from an oak, let alone a spruce from a larch. Flowers are daisies or not-daisies. I guess my instinctive appreciation of how much they might be missing is equivalent to a musician feeling for one who cannot appreciate Mozart. The more you know about nature, the more you see, and this is enriching.

Not that the *experts* always get it right. Possibly the most famous fraud in scientific history was perpetrated upon the Geology Department of the Natural History Museum: Piltdown Man. Between 1908 and 1913 a series of discoveries of fossil bones around the little village of Piltdown in the rural county of Sussex allowed the identification of the "missing link" between apes and humans. Recall that this was before the exploration of Africa as the cradle of human evolution, and that Charles Darwin's observations on likely human descent were at the time almost unsupported by fossil evidence. Here was a discovery devoutly to be wished by the scientific community, and the find of *Eoanthropus dawsoni* fitted the bill. The occupier of the Keeper's position at the time, Arthur Smith Woodward, named the species "in honour of its discoverer" Charles Dawson, a local Sussex solicitor who enjoyed a consider-

The anthropologist and exposer of the Piltdown fraud, Kenneth Oakley, in his working habitat. Oakley is the one in the middle.

able reputation as an antiquary and amateur palaeontologist. In my childhood copy of Arthur Mee's *Children's Encyclopedia* there was a reconstruction of Piltdown Man, as solid as you like, dressed in skins and hammering away with his stone tools. He enjoyed forty years of existence. But in November 1953 he evaporated: the evidence on which he had been based was nothing more than a cheap fraud. Arthur Smith Woodward had been duped by a faker. *Eoanthropus dawsoni* was a confection of dyed bones—"not one of the Piltdown finds genuinely came from Piltdown," as Sir Gavin de Beer, Director of the Museum, wrote in 1955. Piltdown's mandible was probably that of an immature orang-utan. Human skull fragments had been similarly stained with iron to make them "fit." Bone "tools" had been filed down with rasps. Nothing was what it seemed. When Arthur Smith Woodward spent field seasons in the Piltdown pits over several years, he was being set up like any "mark" by a conman. Fortunately for the reputation of the Natural History Museum, its staff were equally involved with the subsequent exposure of the fraud. Kenneth P. Oakley first noticed that the fluorine concentrations of Piltdown bones were far too low to support its supposed antiquity. Once doubt set in, the fraud was obvious—all that staining and filing and the admixture of different sources of bones—so much so that today wonderment is expressed about how a scientist of the stature of Smith Woodward could have fallen for it. There has even been the suggestion that he must have been party to the deception. Given his impeccable record elsewhere, and the fact that he fruitlessly continued "digging" long after the supply of bones had dried up, this is extremely unlikely. He was the patsy, poor man. He *wanted* Piltdown to be genuine and as a consequence was led by the nose.

But the perpetrator must be fingered, and, as in any good whodunnit, there is a wealth of delicious red herrings. Charles Dawson was friendly and shared excavations with the Jesuit anthropologist Marie-Joseph Pierre Teilhard de Chardin, who, for unsurprising reasons, is often referred to just as "Teilhard." By the middle of the twentieth century he was a famous metaphysician whose book *The Phenomenon of Man* was popular, if not required, reading for those with an interest in alternative culture. Although many of Teilhard's ideas have not led to testable hypotheses, his holistic view of the Earth and its place in the cosmos anticipates the stance of many Earth-systems scientists today. He fell seriously foul of the Catholic establishment in Rome, to the extent of having a number of his works proscribed. He also happened to have found the canine tooth of Piltdown Man in 1913. Could he have been responsible for the other fakes? A close examination of the evidence shows that Teilhard was not in the country when certain crucial discoveries were made; the simplest explanation is that he was required as an "expert witness" for the authenticity of material discovered on other occasions. The fraudster was someone else. My favourite but not favoured suggestion for the faker is Sir Arthur Conan Doyle, renowned creator of Sherlock Holmes. Could Piltdown be his invention of the perfect crime, a real-life masterpiece for which all his fiction had been just a rehearsal? The argument in his case is a subtle one. Conan Doyle was a convinced spiritualist, and the then Director of the Natural History Museum, E. Ray Lankester, was a prominent and outspoken sceptic. By discrediting the distinguished Director, the spiritualist camp would have bloodied the nose of the opposition, and made the public sceptical about the sceptic. It is, to say the least, a convoluted argument, although great fun. The evidence linking Conan Doyle with the scene of the crime is very slim, but then I suppose that might be proof of the fiendish cunning of the hoaxer. Conan Doyle as Moriarty, the master criminal! My own view is that this theory represents what might be termed the Jack the Ripper Syndrome: many names have been suggested as the true identity of Jack, but most of them are famous people, like the Prince of Wales or the painter Walter Sickert. Obscurity is much less compelling. Were I an opportunist, I might well be penning a book suggesting Jack the Ripper as the guilty party in the Piltdown fraud.

More modest suggestions for the Piltdown hoaxer include the mammal man at the Natural History Museum, Martin Hinton. He certainly was not enamoured of Arthur Smith Woodward, and may well have experienced more than a modicum of Schadenfreude at his colleague's discomfiture. Hinton is perhaps most justly famed for his important part in persuading the government of the day to eliminate the muskrat, *Ondatra zibethica,* from British shores; this giant rodent had escaped from fur farms and was breeding unchecked, and its burrowing into riverbanks, railway embankments and the like was causing flooding and landslides. The Destructive Imported Animals Act became law in 1932, and allowed for dedicated staff to trap and shoot the invasive pest. The total elimination of the muskrat was achieved by 1939. Hinton should have been recognized by a grateful nation, but he never was. "Muskrat Disaster Averted" is not a likely headline. The same escape story was repeated several decades later with the coypu, *Myocastor coypus,* as principal actor, and I can remember the little white vans belonging to the coypu exterminators lurking around the marshes in East Anglia. Hinton's relevance to the Piltdown Man scandal was the discovery, many years after his death, of a canvas trunk apparently belonging to him hidden away in an obscure attic of the Museum in which were found various bits of bone and stains. It could have been a rehearsal for the fraud. However, it is clear that Hinton, who was a young temporary assistant at the time of the Piltdown "discoveries," was never at the pits during the excavations. It seems much more likely that the trunk contained his, or someone else's, experimental attempts to

The examination of the Piltdown skull, from a painting in the Geological Society of London. Dawson is the figure whose face is closest to the portrait of Charles Darwin. E. Ray Lankester, the Director of the Natural History Museum, is farthest to the right.

replicate the techniques of the fraudster, something that anyone interested in mammal bones would have wanted to try out once the hoax emerged. Perhaps the most surprising thing about the story is the fact that a large trunk could be hidden away unopened for years in the Museum. However, there is a tradition of chaos and concealment among the mammal workers. The disorder in the office of Hinton's successor, Arthur Hopwood, was legendary. Tony Sutcliffe, who followed on from Hopwood as fossil mammal man, told me that he had to help with the dismemberment of the heaps after Hopwood's death in 1961. A curious, musty smell pervaded the midden. Underneath piles of unanswered letters and scientific reports was a shoe box tied up with string. Inside the box was a brace of mummified pheasants and a label carrying the message: "A small gift to Dr. Hopwood, Christmas 1958."

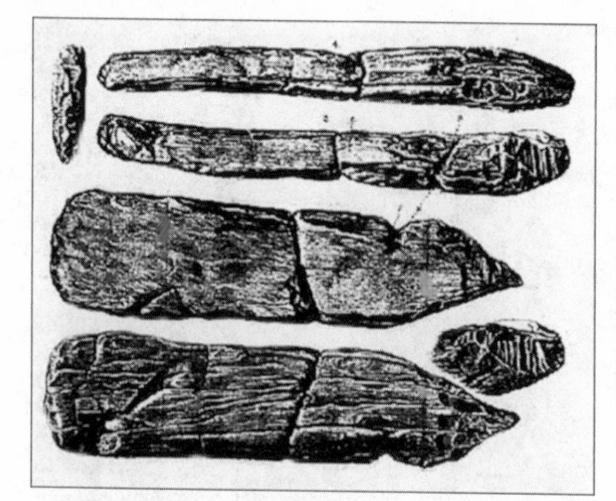

Some of the artefacts associated with the Piltdown site. *Left*, the notorious "cricket bat"; *below*, stone tools published in the *Quarterly Journal of the Geological Society of London* in 1914

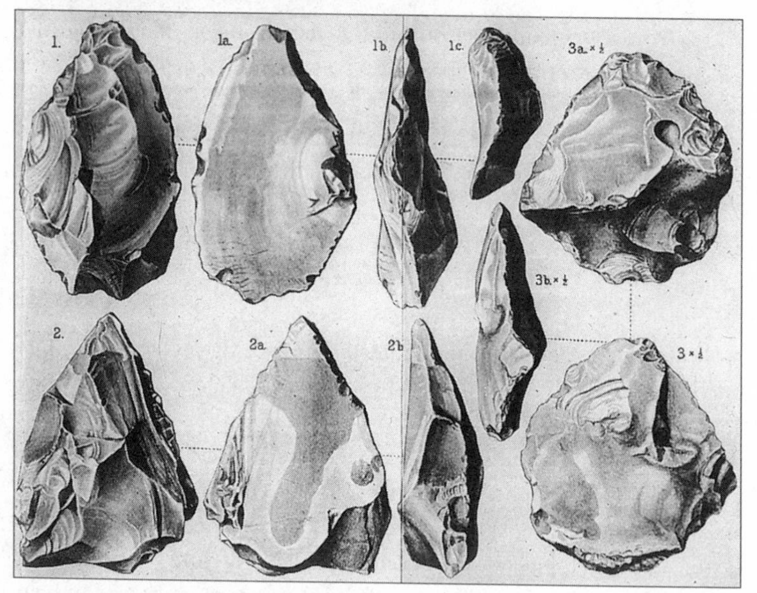

In fact, the guilty party is now known beyond reasonable doubt: it was Charles Dawson, the eponym of Mr. Piltdown himself. Miles Russell has shown in *Piltdown Man: The Secret Life of Charles Dawson* that Dawson was a serious, serial fraudster. Piltdown Man was merely his most ambitious wheeze. Dawson had produced a number of archaeological objects that were designed to increase his reputation and "solve" archaeological problems, or perhaps provoke gasps of admiring surprise from the professionals. He found tiles to "prove" the Roman name of Pevensey Castle, or he discovered unique Bronze Age artefacts—which were probably fakes. To the antiquarian world there seemed no end to his luck. Many of his discoveries are not readily authenticated, but neither are they necessarily bogus. Infuriatingly, though, at least some of his discoveries *are* genuine, and his published historical compilations are undoubtedly useful, so he was probably a talented archaeologist who attempted to puff himself up beyond his true worth. He simply could not resist self-aggrandizement; and his duplicity went to his grave undetected. His wife always felt he had failed to say something important to her before he died, but it seems unlikely that it would have been a confession. In his photographs Dawson looks every inch the Edwardian gent: erect and moustachioed, medals on his chest, a veritable pillar of society. He must have been entirely plausible in person, skilled at "playing" Smith Woodward, who was led inexorably on through the Piltdown years by a discovery here, a discovery there, all of them noteworthy. Dawson was also, in retrospect, the obvious person to perpetrate the fraud. After all, he found the objects, or was close by when others did so. He discovered the new sites. He orchestrated the whole project. To use the hoary criteria of the fictional detective, he had both the motive and the opportunity. He had even been discovered by one of his visitors furtively dyeing bits of bone. Indeed, a fellow amateur archaeologist, Harry Morris, had entertained private suspicions about his Sussex contemporary from the first, but had kept his counsel for reasons of his own. Dawson may be a less glamorous solution to the mystery of the Piltdown forger than Teilhard or Conan Doyle, but he is almost certainly the correct culprit. His memorial, a sandstone monolith, still stands in the grounds of Barkham Manor, near the pit that was once world famous. Like the pit, the stone is now overgrown and unacknowledged, but its inscription can still be made out: "Here in the old river gravel Mr. Charles Dawson FSA found the fossil skull of Piltdown Man, 1912–1913." Maybe its complete obliteration would be most appropriate for a man who diverted history for forty years.

The anthropologists in the Natural History Museum have long

since outlived the scandal. Kenneth Oakley went on to a distinguished career, and published widely on hominid artefacts and remains until he became tragically disabled with multiple sclerosis—even then, he continued to struggle into the Department of Palaeontology on crutches. The techniques he pioneered were developed further by another of the Museum's formidable and able women researchers, Theya Molleson, who became the expert on ancient pathologies reflected in bones, or diet revealed by wear patterns on teeth. Meanwhile, discoveries in Africa were rewriting the story of the emergence of humankind; far from originating on the Sussex downs, *Homo sapiens* arose in Africa about a hundred thousand years ago. During my lifetime in the Museum I have seen this idea change from one of several theories to become something close to an accepted fact. Dawson's bogus "missing link" has been replaced by a whole gallery of fossil species, some on the line to modern humans, others on interesting side branches. Mankind in the widest sense moved out of Africa in several pulses, migrating in response to climate change and ecological opportunities. He did indeed reach the British Isles—the story to which Dawson might have contributed if he had not had such vainglorious ambitions. There is currently a programme of research of which the Museum is an important part that goes under the acronym of AHOB—the ancient human occupation of Britain. In overall charge of the project is Dr. Christopher Stringer. The money for it comes from the Leverhulme Trust, one of those charitable organizations without whose patronage it would be difficult to do large-scale science at all. Lord Leverhulme made his fortune from soap products in the early part of the twentieth century; until his death at a considerable age, Leverhulme would preside at an annual party of the Trust's beneficiaries from a kind of throne. Many scientists—including this one—have reason to be grateful to him.

It is one of the more ironic aspects of the investigation of early human occupation of Britain that one of the most important sites, Boxgrove, lies in the county of Sussex, where Dawson had caused such a time-wasting diversion. There was a real history just waiting to be found. Around an ancient water hole five hundred thousand years ago rhinoceros, horse and deer came to drink, and human hunters were waiting for them. Archaeologists have found splendidly preserved

Palaeolithic flint tools, accompanied by the flakes wasted during their manufacture, together with clearly butchered bones, some cracked to extract the nutritious bone marrow. Mark Roberts, who directed the excavations at Boxgrove, has pointed out that these discoveries are consistent with the notion of "man the hunter" rather than his being an opportunistic scavenger as others had claimed. Killing skills root deep into our history. As to the identity of the killers, the discovery of a limb bone and other fragments indicated to Chris Stringer that he was *Homo heidelbergensis,* one of our predecessor species, but one that was human in many aspects of his behaviour such as hunting co-operatively. As recently as 2005, even earlier evidence of human occupation, up to seven hundred thousand years old, was discovered in Pakefield, on the coast in the East Anglian county of Suffolk. This is at present the earliest occurrence of humans north of the Alps. The crumbling cliffs of Suffolk have yielded many fossils of animals in the past. I have even picked up heavy bits of dark mammoth bone myself on the bleak shores. It is a lonely place to work, and real dedication is required to survey every new piece of sand or clay dislodged by the unforgiving onslaught of the North Sea. The Suffolk coast in winter is frequented by two kinds of people, both of whose sanity might be questioned by the population at large: onshore fishermen and palaeontologists. What they have in common is oilskins, an infinite capacity for hope and a certain camaraderie tinged with competitiveness. The new discovery was a tribute to the persistence and powers of observation of two local amateur collectors, Paul Durbidge and Bob Mutch, who discovered some of the crucial evidence. Tools worked by human hands were discovered in situ among the bones of rhino and spotted hyena. The tools are not much like the exquisitely wrought artefacts of later "stone age" cultures, but the freshly broken flint surfaces are unmistakable artefacts to the trained eye. What eludes the investigators so far are the critical bones of the earlier British *Homo* himself.

The AHOB project is revealing Britain as one of the best places to study human history in northern Europe. Wherever new discoveries in the field are made, the materials recovered are still what they always were: fossils, artefacts and sediments. It is still necessary to get the hands dirty and to muddy the seats of trousers. Nowadays, much more attention is paid to tiny fossils of pollen grains and vole teeth than was once the case, but the importance of taxonomic expertise remains the same—you need people who know their stuff: Adrian Lister for the elephants and their relatives, Andy Currant for the voles, Russell Coope for the beetles. What is thoroughly new is the variety of techniques that can be applied to the material collected. No tomfoolery with specimens is possible now that they can be chemically analysed down to their very atoms, and their geochronology investigated using Radiocarbon dates (if they are not too old), Mass Spectrometric Uranium series, Electron Spin Resonance and Optically Stimulated Luminescence. This is white-coated laboratory science: "real" science, doubtless, to devotees of gadgets and gizmos. One of the intentions is to use oxygen isotope measurements from fossils to link the sporadic fossil record in Europe with the standard climatic cycles recognized from sampling continuously deposited oceanic sediments; ratios of these isotopes fluctuate in harmony with the ambient temperature. This provides crucial information, because early human history coincides with the climatic fluctuations of the Pleistocene, a period which began 1.8 million years ago and lasted until twelve thousand years before the present. This is a large part of the "ice age"—in fact, it comprises a whole series of advances and subsequent retreats of the northerly ice sheet.* As a species, we always have been in thrall to climate change, as we may well be again over the next century if global warming predictions are fulfilled. What AHOB has proved beyond doubt is that mankind's history in Britain is one of repopulation with long intervening periods when apparently no people were able to cope with the frigid conditions. They retreated southwards into continental Europe. The longest of these periods was from 180,000 to 60,000 years ago, after which humans, including Neanderthals, recolonized when the climate ameliorated. Another, younger frigid period between 22,000 and 16,000 years ago has also so far failed to yield any evidence of human occupation of Britain. Since this period included the Last Glacial Maximum, when ice extended south almost to the

*These are calibrated against the so-called Marine Isotope Stages, which now provide a unified timescale for the ice ages. Tapping into this standardized clock is not a simple matter, however, and much argument still takes place over the exact age of particular sites. The succession of mammal species is important independent evidence.

Home Counties, this fact may not be altogether surprising. Only after this time were our own distant ancestors, belonging to *Homo sapiens*, able to move in. If we had a notion that inhabitants of these islands were invaded by Vikings and Normans but otherwise inviolate, then it is clearly wrong. Repeated invasion has been the very stuff of our history.

The image of the modern laboratory is one of humming machines counting atoms of oxygen isotopes while a scientific technician in a white coat wearing thick glasses and thin gloves stares at a computer console. In my experience this image is almost entirely correct. Hence I am grateful for the existence of the laboratory, or, as I now have to call it, the Conservation Unit, in the Palaeontology Department. It still has fossil bones and microscopes and interesting things on benches. People still get their hands dirty with ordinary dirt. Much of its business these days is preparing casts—including some famous hominid specimens—which can be made with an accuracy and stained to resemble the original in a way that Charles Dawson would surely have admired. Here the "BM *Archaeopteryx*" is copied for other museums around the world, prepared in a light resin that imitates the original in a way far more delicate than the old plaster-of-Paris casts. The laboratory is in the basement of the building, so that large lumps of rock can be admitted directly through the double doors at the back. From time to time something so important is discovered that everybody drops everything to get it in through those doors. The discovery of the dinosaur *Baryonyx* was one such occasion. Mr. Walker found the claw of *Baryonyx* in 1983 in a quarry in Cretaceous rocks at Otley in the English Weald. From a subsequent visit to the site by Museum scientists, it became clear that much of the skeleton was preserved in sandstone blocks in the quarry; furthermore, it seemed to be an altogether new kind of dinosaur. This kind of excitement is rare. All the available staff took off to Otley to collect as much as possible, carefully recording the relative position of bones and making plaster cradles for more fragile specimens. When the dinosaur was eventually brought in through the back doors, it was the job of master preparator Ron Croucher to extract it from the rock. This was a task somewhat similar in magnitude to carving the faces of those presidents of the United States of America on Mount Rushmore, except that close-up details matter rather more, and it was a question of extracting reality from the rock rather than sculpting the rock to a known design.

Ron Croucher is a most self-effacing kind of man, but I believe he has little doubt that *Baryonyx walkeri* is his masterpiece. His apprenticeship as a preparator went back to the early days of the Palaeontology Department. In the old Museum the "lab" was housed in dreadful conditions at the north side of the building. It was so cold in winter that a series of electric fires were propped all over the place to keep the staff warm, and the lighting was so poor that it was sometimes necessary to wait for the sun to creep into the room before a particularly delicate task could be undertaken. Nonetheless, innovative work was carried out there. The staff pioneered the use of resins in casting fossils, which is now routine. On one occasion in the old hierarchical days, a very senior scientist and mollusc specialist, Dr. L. R. Cox, appeared in the laboratory at teatime in search of milk; apparently, his tea club of senior chaps had run out of it. A beaker of the latest resin lay on the bench, opaque and white. Without a by-your-leave, Dr. Cox appropriated the liquor and swept out. The subsequent reaction in the senior tea club is not recorded. Ron Croucher's predecessor, Arthur Rixon, pioneered the extraction of fishes from limestone by acid solution. A fish—like us—has bones of calcium phosphate, whereas limestone is calcium carbonate; the latter dissolves in certain acids, acetic acid perhaps being most commonly employed, leaving the bones untouched. The result can be the most exquisite preparations of fossils, as precise as a dissection, with no tool brought to bear on the fragile skeleton.

No such short cuts were possible in the case of *Baryonyx.* Ron had to cut off the sandstone surrounding the bones grain by grain using a vibrating air pen with a hardened tip. It is painstaking work: a moment's lapse in concentration can polish off the work of days. In the case of *Baryonyx,* extraction was even more difficult, because the bones were softer than the surrounding rock. Certainly much more than a year of Ron's life went into the work; he says that he lost all sense of time after a while. The final result is a reconstruction of the dinosaur as a specialized fish eater, a predator with an almost crocodile-like skull and huge claws; it is related to *Spinosaurus,* but tells a new story about the adaptability of the ruling reptiles to different life habits. Ron's contribu-

Preparing a mounted skeleton of the dinosaur *Baryonyx* in the "laboratory"

tion was acknowledged in the scientific description of *Baryonyx* by Alan Charig and Angela Milner—a formal nod towards his hours and hours of patient toil. When dinosaurs appear in books and films as realistic as if they had been plucked from the Cretaceous by a time machine, it is easy to forget that everything we know is the result of the labours of unsung masters like Ron Croucher. Reality is extracted out of sight of the public in back rooms full of half-exposed bones.

Some people think of huge dinosaurs as more or less synonymous with fossils. At the other extreme are the molecules making up the genes that have controlled the course of evolution from microbe to mankind. It might be thought that fossils and genes would never meet, but recent research has made it happen. Palaeobotanists study the fossils of plants. More primitive plants have survived to the present day than animals, so you can find flourishing examples of *Ginkgo* trees, cycads or monkey puzzle trees that would not have been out of place when *Baryonyx* walked abroad in the Weald. There have been a number of palaeobotanists associated with the Natural History Museum, not all of them happy appointments, although the collections are as vast as you might expect. K. I. M. Chesters, who was palaeobotanist when I joined the Museum, seemed rather uninterested in fossil plants, which is not a good qualification for the job, although she had produced some publications in the 1950s: she had a large loom in her office, and I believe she was weaving when she should have been working. When she married another member of the department, the "alga man" Graham Elliott, wags suggested he had been bribed by the Keeper to do the right thing for the department. She left shortly afterwards.

The present incumbent, Paul Kenrick, is a humorous and dynamic man in his forties who is concerned particularly with the early history of land plants, and knows a lot about those ancient groups, like ferns and horsetails, that have survived from what is often referred to as "deep time." Fossils provide a direct way of learning about early plants and their relationships. Another route is to study the genomes of the survivors from former times. Similarities and differences in their gene sequence patterns reveal their degrees of relatedness—at least, if the right "designer gene" can be identified, as I described in the last chapter. This method has led to the recognition of evolutionary branching events that happened tens or even hundreds of millions of years ago. It may also suggest in turn where a particular fossil may fit on the evolutionary tree, thus allowing cross-fertilization between palaeontology and molecular biology. As a high point of this method, a remarkable, sequence-based evolutionary tree for all the flowering plants was produced in 1993, something that botanists had desired, but failed, to do since Darwin's time. Since the early seed-bearing plants were contemporaries of the dinosaurs, this was like being provided with a telescope capable of looking back a hundred million years.

Paul Kenrick and his colleague have been looking at a group of survivors called lycopsids. In the coal swamps of the Carboniferous three hundred million years ago lycopsids grew into huge trees—their bark and roots are common fossils found in association with coal seams. Museum drawers are stacked with examples. Compared with the flowering plants, lycopsids are Methuselahs. Three surviving families are all that remain of this once great group, but these had already diverged from one another back in the Carboniferous. One of these families is the Lycopodiacea, of which only herbs survive today, belonging to three genera. One genus, *Huperzia,* is an epiphyte that lives its life attached to the branches of tropical trees. The other two, *Lycopodium* and *Lycopodiella,* are the club mosses, familiar low herbs on damp heaths and moors. It would clearly be interesting to know whether species of these genera are *all* ancient survivors, or whether they evolved more recently, perhaps alongside the flowering plants. The living species were investigated using sequences from the gene *rbcL,* which had previously proved so useful for the flowering plants. Then the sequence differences between lineages were used as a kind of clock to estimate the time of divergence—that is, when they separated into distinct evolutionary lines. It was discovered that *Huperzia* clearly divided into two groups of species: one Neotropical in South America, the other Palaeotropical on the opposite side of the Atlantic Ocean. Furthermore, the diversification of the two groups happened in the Cretaceous, more than seventy million years ago. This is consistent with the widening of the Atlantic Ocean following the break-up of the ancient continent of Gondwana, which was the southern part of the still greater supercontinent, Pangaea. Many other plants have a similar tale to tell—evolution took on distinct trajectories once the old and new worlds had separated. By contrast, much of the evolution of *Lycopodium* and *Lycopodiella* species happened in much more recent times in the late Tertiary, different species being found in different and separate localities. So it seems that, despite their antiquity as a family, lycopsid evolution did not simply sit still. New species groups arose in response to events in geology. On the other hand, the separation into different genera probably happened as far back as the Permian 260 million years ago, and so *all* the lycopsids are, in another sense, living fossils. It remains true that, when one looks at these plants, one is looking back hundreds of millions of years into the geological past. The differences between species are relatively

Living lycopsids in New Zealand, showing a succession of vertical sporophytes, looking like little "brushes"

small—a matter for the expert—and the living species evidently arose as a result of geographical separation. This example shows how study of humble herbs can marry geology, fossils and molecules in a most stimulating and satisfying way. I can't help wondering what the Great Men like Spath would have thought of it all.

There is no danger of running out of new discoveries of fossils. I have been amazed by what has been prised from the rocks in China. As well as feathered dinosaurs and fossil flowers more ancient than any known previously, tiny fossilized embryos have been discovered in rocks still older than the Cambrian, the first direct evidence for animals in strata of this age. Seek hard enough, and finds shall be made; the book of the history of life must be continually rewritten. It could be argued that the recognition and naming of fossil species are less urgent than those of living animals and plants. They are not going anywhere from their sedimentary tombs—and, after all, most fossils belong to species already long extinct. This is to miss the point, for we are what history has made us. There are lessons to be drawn from the past extinctions and climatic crises that life has weathered—histories that may well equip us to cope better with the climate changes that are happening now. For example, the climatic warming of today marks a return to conditions on the "greenhouse earth" of mid-Tertiary times: suddenly the expertise of a palaeontologist who knows about this period will seem less arcane. Equally, the impoverished world that followed upon the great mass extinction at the end of the Permian period 248 million years ago provides a chilling prospect for humankind if we continue to degrade our planet in the way that we have been. This provides a pragmatic, sensible, utilitarian reason for studying the past, but one that is anthropocentric at root. I don't want to reduce the fossil record to "lessons from history." The real joy of discovery is to see the exuberance of life, those trilobites with tridents, or those great flying reptiles as big as gliders—organisms almost as exotic as the confabulations of Hieronymus Bosch, but thriving here on Earth long ago. To know about the wonderful excursions that life has taken is to be enriched, to be made aware of the fecundity of our small planet. This is what motivates palaeontologists to tap away for weeks on end at ungrateful rocks. This is the point of securing the booty recovered as testimony for generations of scientists to come in the drawers of the hidden museum.

4

Animalia

Zoology embraces all animals, not just the appealing furry or feathered varieties, so I will start with snails. Slugs and snails are regarded by some people as the least appealing of nature's productions. As the nursery rhyme tells us, little boys are made of them, with added puppy dogs' tails, to contrast with all that *sugar and spice and all things nice* that little girls are made of. But the Class Gastropoda is also one of the most diverse outside the insects and includes many miraculously beautiful coiled shells. One of the least spectacular, if most familiar, among the seashells are the little turreted winkles that one finds clustered on rocky shores. Sometimes they are found in such profusion as to make walking over the rocks a difficult business. Winkles are edible, although hard work. They are sold cooked in small tubs, and a traditional seaside delight was to "winkle" out the tough little morsels with a pin—hence the name of the pointy winklepicker shoes that enjoyed a brief heyday in the 1950s. A slightly more refined name for the little mollusc is periwinkle, but that is also the moniker of a plant, so I prefer the shorter version. David Reid in the Natural History Museum is devoted to winkles, and knows more about them than anyone. He is tall and thin, in early middle age, with a short, well-cared-for beard and glittering, intense eyes. David becomes extraordinarily animated when he talks about his favourite snails. He is to be found in the basement of the old building not far from the site of my first office. Almost all the other zoologists have moved to the new Darwin Centre, but the molluscs are still where they have always been. Such permanence is rather comforting, so when I visited David I could briefly imagine that I was still the young tyro strolling down the wide corridor with my life ahead of me. Winkles used to be placed in the single genus *Littorina,* the common (peri)winkle being *Littorina littorea,* which tells you all you need to know about its littoral—that is, shoreline—habitat. Nowadays, the winkles are divided into a number of different genera. "The great thing about them," David enthuses, "is that you can collect them all over the world so easily. There is probably no better group to use to examine the relationship between geography and species." Winkles do not run away when the scientist tries to catch them, they do not fall to pieces, nor do they have Threatened or Vulnerable status on conservation lists. Their flesh can be used to extract DNA readily enough. They are ideal subjects to winkle out the truth about evolution.

David Reid gathered together everything that was then known about *Littorina* in a huge monograph published by the Ray Society* in 1996, all 463 pages of it. The book is monumental. David tells me with characteristic self-deprecation that when it was published a Scandinavian colleague with less than perfect English wrote to him congratulating him on producing "a major millstone." There must be hundreds of photographs of shells, and nearly as many of the radulae—the rasp-like "teeth" of molluscs—taken with the electron microscope, to say nothing of dozens of drawings of the penis, which is a critical character in winkles. There are details of the sculpture on the shells to help the reader identify a species in hand. It turns out that not only do winkles vary between species, but they also offer excellent examples of variation *within* species according to differences in habitat; winkles nestled among the barnacles will have different shells from those on more exposed rocks. In some localities there are a number of separate species

*John Ray (1628–1705) was a pioneer botanist who has been somewhat, and unjustly, overshadowed by Linnaeus. His name is commemorated in a series of learned and beautifully produced monographs of the Ray Society, of which David Reid's is the 164th.

on a single shore adapted to different heights above sea level. There are knobbly ones, ridged ones and nearly smooth ones. Some winkles have planktonic larvae, while others do not. In some species there is a good reason for a particular feature. For example, winkles living at high latitudes have the outer shell layer made of the form of calcium carbonate known as calcite, whereas the tropical ones tend to have the same layer made of its other form, aragonite. Calcite is less soluble in cold water than aragonite, so it makes sense to use this material in more arctic climates. Some species of winkle have to live on seaweeds; others refuse to do so. As more is learned it becomes clearer just why there are so many species of this lowly snail, and how they manage to coexist.

It might be thought that after the production of such a work there would be nothing more to be said. But David Reid believes that the truth about winkles is emerging only now that studies of molecular sequences have been added to the equation. In the last ten years he has been working with S. T. Williams on tropical winkles. The phylogenetic trees produced from the sequence data revealed that species from Australia and Africa were evolutionarily quite separate—so much so that they might be separated into different genera, called *Austrolittorina* and *Afrolittorina*, respectively, for obvious reasons. The most startling fact I learned about these two genera is that the nine species now recognized as belonging to them were formerly thought of as but a single, highly variable species. Furthermore, there are species of *Austrolittorina* and *Afrolittorina* that are completely indistinguishable on the shape and ornament of their shells. Yet their molecules tell us that they had separate evolutionary origins. The shells alone could not reveal this story. There could scarcely be a more powerful example of cryptic species—their recognition would have been impossible before the "molecular age." The similar demands of shore habitats in their widely separated localities determined that the winkle shells evolved into an identically similar shape. The arrangement of the continents and islands that allowed the species to evolve in separation was itself a consequence of geological processes, as the tectonic plates carry on their unseen business, in this case, the separation of Australia from Africa as their respective plates bore them away. So geology, and evolution, and habitat combine to specify what a winkle should look like. The world is a lot

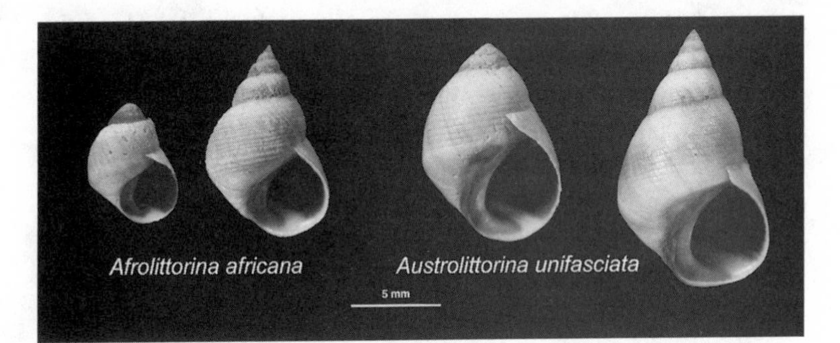

Sea shells can look very similar but have separate evolutionary origins: the "winkles" *Afrolittorina* and *Austrolittorina*.

richer than one might imagine. The apparently esoteric activity of studying marine snails tells us much about how this richness came to be. Consider the snail and be wise.

A little farther along the basement corridor, John Taylor still works in the same office he walked into in 1965; it retains the richly patinated hardwood cupboards with the names of molluscan families on the doors, so like those in my own first office. His office is a cramped space at the western end of the old building, with low ceilings revealing the iron skeleton of the Museum, beams and bolts and all. Visitors have to sidle alongside cabinets to get to the inner sanctum where the computer and the scientist reside. John retired some years ago, but continues to come into work almost every day. This might surprise some people for whom work has been an obligation and a chore, and for whom working for nothing might seem a strange concept, but then after forty years John Taylor is still as fascinated as ever by his clams—his work is simply who he is. Besides, there is always something new and exciting to discover. Just at the moment, he is researching into clams that harbour colourless sulphur bacteria in their tissues. These clams are specialists in the molluscan world: they actually cultivate the bacteria in their gills, and many species are adapted to absorb nutrients directly from the "plants" in their anatomical garden. John and his colleagues have shown that this adaptation is widespread in many different habitats: in mangrove swamps, or on seafloors low in oxygen, or around those strange, deep-sea sulphurous vents that exhale hot water along the mid-ocean ridges. The clams get food, and the bacteria grow in a protected environment—indeed, some clams have developed behaviours that help to "feed" the bacteria with the sulphur they require. Both parties benefit. It is a classic case of symbiosis. Some of these special clams can live in marine habitats that are very low in oxygen, which actually helps some bacteria thrive. It is a marvellous demonstration of the vigour of life to colonize even the most unpromising environments. And it is possible that such apparently obscure species will be important in coping with the polluted seas to which we humans are contributing. Taylor has proved that this specialist life habit has evolved no fewer than five times independently in the clams, but especially in a group called lucinoids. Once again, new evidence from DNA helps to prove this. It seems like any trick good enough to earn a livelihood in nature is good enough to be copied. And again we run into the taxonomic underpinning of the Museum, because John and Emily, his partner, have discovered a host of new species of symbiotic clams, even in places that have been sampled several times before. Some of them are tiny, and you could imagine how they would be overlooked. One or two are quite large, and one, from the California coast, is a really substantial fistful of an animal, which John first recognized from a museum collection. They need names, of course, and once described and christened will add another small turret to the edifice of biodiversity. That's how enduring knowledge of the natural world grows: little by little with the help of enthusiasts.

It is not surprising that the rich attractions of the mollusc collection have made it the target of unscrupulous collectors. There is a kind of disease that afflicts certain naturalists that gives them an irresistible compulsion to own their objects of study as well as understand them. Sometimes this results in collections of lasting value to science—I have already mentioned a number of these coming as bequests to the Museum. In other cases the collecting compulsion takes a pathological turn—they *have* to own an example of a desired species, whatever it takes. The shell collector Tom Paine was one of these. As a trusted amateur expert he was allowed access to the collections for many years. Eventually, staff became worried by a series of disappearances from the

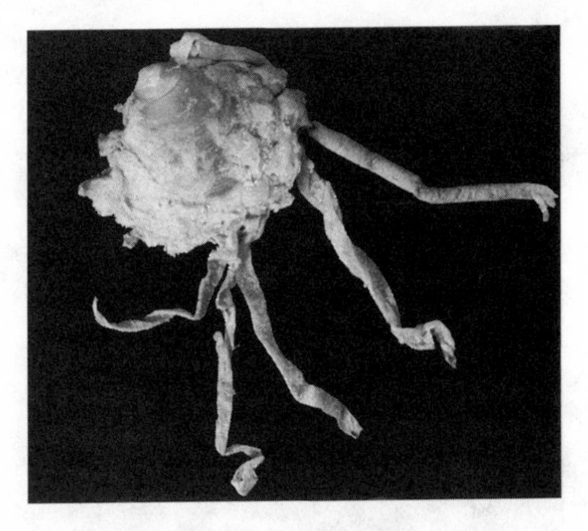

One of John Taylor's discoveries in Western Australia, which was named by him and Glover in 1997; a lucinid clam *Rasta thiophila* (thiophila means "sulphur lover")

drawers that seemed to follow upon his visits. These absences even included some of the precious type specimens—not the holotype maybe, but some of the other specimens that were part of the type collection. He could never be caught red-handed, but he was eventually excluded from access to the London collections. In the meantime he had apparently made similar light-fingered visits to other museums all around the world. After his death in the 1990s he bequeathed a magnificent and enormous collection to the National Museum of Wales, but the origin of many of the specimens will probably never be sorted out. An even more blatant example was another visitor who raided the birds' eggs. This is one of the more familiar obsessions, and one that I can understand, because as a young boy, before it became illegal, I briefly collected eggs myself, as had my countryman father before me. I am ashamed of it now, but I can recall the excitement of coming across an uncommon species; and birds' eggs are fragile and beautiful objects. They are just the kind of thing to attract an oddball kleptomaniac. This particular thief appeared in a wheelchair; he secreted the desired birds' eggs inside women's tights that he then tucked into his trousers before making good his exit. He was discovered and prosecuted. It later transpired that the reason he was in a wheelchair in the first place was that he had had an accident stealing copper from electrical installations. Shells, eggs, fossils—it matters not—all intrusions into the collections

Another beautiful new clam from Western Australia, *Plicolucina flabellata*

are taken very seriously, and have to be reported to the Trustees by the Keeper of the department concerned.

Perhaps I should explain about the Trustees. The Director answers only to the Trustees, and the Trustees answer only to God—or at least to the government of the day. They are the ultimate governing body of the Museum. They have been there from the inception of the British Museum at Bloomsbury. Sir Hans Sloane, whose collection formed the basis of the British Museum, made it clear in his will that his collections were to be entrusted to Trustees, and that they, in turn, should offer them to the King for the nation. The original list of Trustees numbered no fewer than fifty-one, most of them friends or relations of Sloane himself, and most of them beyond reproach, for Sloane was doubtless aware of the corruption pervading society in the first half of the eighteenth century. He realized that a philanthropist needed people to look out for his interests, lest things "go astray." Even so, it was a good idea to have plenty of them, just in case one of them got any irregular ideas. Thanks to a set of influential Trustees, the "Act for the purchase of the Museum . . . of Sir Hans Sloane . . . &c" received the Royal Assent on 7 June 1753. The result (eventually) was a building to house the collections, and a set of Keepers to look after them. The burgeoning British Museum has been a drain on the resources of government ever since, and no doubt there have been civil servants who have scrutinized the

The monument to Sir Hans Sloane in the
Chelsea Physic Garden c. 1910

accounts and wondered whether a little judicious pruning might actually save a few pounds.

When the Natural History Museum split off from its parent body, another set of twelve Trustees were appointed. These worthies still meet a few times a year, at which time the Director has to present his policies and resolutions for approval. The word of the Trustees is still passed down to the shop floor with all the gravity of holy writ being handed down on tablets of stone.

For most of the scientists working on beetles, beans or *Brachiosaurus,* the Trustees might as well have been living on a distant planet for all that was seen of them. The regular changes in Trustees didn't seem to make much difference to our daily lives. They tended to be distinguished clerics, academics or aristocrats with a natural history bent. What they did was all happening "upstairs" somewhere; downstairs, life went on. Very rarely, a visit from one of the Trustees was scheduled, and this entailed desperately sorting through the heaps that littered the office, returning books to the library, putting specimens away and generally scrubbing up. The office was always spotless by the time the Trustee arrived, or the Keeper would want to know the reason why.

As with so many aspects of British life, Mrs. Thatcher transformed the way the Museum worked. In the 1980s the composition of Trustees changed. Now it was deemed appropriate to have successful businesspersons as a sizeable proportion of the Trustees; out went bishops and the retired Sibthorpian Professor from Oxford University, and in came the Chief Executive Officers. This was part and parcel of instilling a new spirit of realism into our ivory towers, of shaking up the old Civil Service by making it conform to the business paradigm that was considered the model for the successful running of any organization. Black Rolls-Royces pulled up on Trustees' days as the Head of Whatever plc made a slot in his busy life to oversee the policies of a great museum full of butterflies, trilobites and exhibitions. A museum is evidently just the thing for a captain of industry to knock into shape. In fact, many of the new Trustees did not do a bad job, but it soon became difficult to regard them as exemplars of the Great and Good in the tried and tested sense. Among the appointees was Gerald Ronson, a very successful businessman and head of the Heron Group. I am not sure that Mr. Ronson necessarily had any interest in natural history before becoming a Trustee of the Museum. I met him on one occasion when he did the rounds of our department, and I remember a very fine, pinstriped double-breasted suit that seemed to fill the room, while I attempted to explain why trilobites were important. I was regarded with the sceptical air that might otherwise be reserved for a salesman attempting to flog a second-hand Jag. Clearly, we approached one another from different worlds, the distinction between us being principally that, while I had absolutely no influence on his world, potentially he might have much influence on mine.

However, he did help considerably in part-financing the new dinosaur galleries in the west end of the Natural History Museum, our most popular attraction, and he was instrumental in securing favourable terms for the Museum's acquisition of a former bus depot in Wandsworth, South London. This building has been essential as an overflow store, and without it the reorganization of the collections during the opening of the new Darwin Centre would have been impossible. As this is written, its financial potential is being realized to help fund the second phase of the Darwin Centre. It would be difficult to argue against the proposition that Mr. Ronson's contributions to the Natural History Museum have exceeded those of the average archbishop or aristocrat. Presumably, had all gone smoothly his good works would have registered in the nebulous and mysterious honours system for which Britain is justly renowned. However, in 1986, Ronson was embroiled in allegations of irregularities over the takeover of the Distillers Group by the mighty brewing company Guinness. He and his associates had attempted to manipulate the stock market to inflate the price of Guinness shares. The case came to be known as the Guinness Scandal, and revealed the unacceptable face of Thatcherism whereby the rich and powerful were capable of controlling outcomes for profit. In 1990, Gerald Ronson was "sent down," along with his collaborator Ernest Saunders; the latter was released from prison because of his Alzheimer's disease—from which, judging by the vigorous defence of his actions he made subsequently, he was the first person ever to recover. The whole affair left an unpleasant aftertaste.

It also introduced ambiguity into the role of the Trustee. Trustees were by *definition* trustworthy: they were supposedly incapable of acts that fell below the probity expected of this role in public life. The idea that any Trustee could be a jailbird would have set Sir Richard Owen or Sir Gavin de Beer spinning in their respective graves. Trustees become important at certain times: for example, they oversee the appointment of new Directors, and they must approve any major policy shifts. It can be a thankless task working *pro bono publicum.* By and large, the Trustees I know are high-minded people who are genuinely interested in contributing their wisdom to a national institution. However, they may not always understand the human cost of their decisions. In 1990, when Neil Chalmers as Director restructured the Museum, the staff discovered that they were actually employees of the Trustees, rather than of the Civil Service. This meant that the Trustees could declare "areas of redundancy"; in other words, they could give people the sack. This came as something of a shock to those who had joined the organization when it was just a remote limb of the Civil Service, and believed that the only grounds for dismissal were "persistent and gross moral turpitude." Idleness and inappropriate behaviour were thought of as only venial, rather than mortal, sins. As we have seen with Leslie Bairstow, a life of non-productivity had never been an obstacle to tenure in the past. Now, the rules were all changed, and the Trustees could give the nod to survival or departure. Each department on the science side had to identify "areas of redundancy" as part of the restructuring. It was a ghastly time.

To give one example, Dr. Alan Gentry, the mammal specialist, was forced to retire early. Alan's field of expertise may seem obscure: he is an authority on fossil bovids. He is one of the few people in the world who can identify antelopes from their limb bones. Alan himself could win a Nobel Prize for Diffidence if there were one and one might understand why he should seem so vulnerable. But as so often in the Museum, appearances are deceptive. Remember that the reason why our distant ancestors came down from the trees in Africa may have been because the climate became more arid. And one of the best ways to recognize what was happening in the distant past is to study the fossil bones of bovids, which are a hundred times commoner than fossils of early humans. In fact, Alan Gentry's research has proved pivotal in obtaining the full picture of the changing African ecology. He has received collections from all over the continent; his expertise is in demand. It is a measure of Alan's devotion to his field that he has continued to work for nothing ever since, even after having been shown the door in unpleasant circumstances.

The Chairman of the Trustees at this time was Sir Walter Bodmer, renowned for his work for what was then known as the Imperial Cancer Research Fund, now part of Cancer Research UK. In conjunction with Neil Chalmers, he was determined to modernize the science departments. He was something of a tough guy, insisting throughout the traumas that the management had a right to manage, which is rather like saying a dictator has a right to dictate. At almost the same time, the science departments were being dethroned from their traditionally central role. Rather than their being represented in Museum management by the Keeper from every department, a Science Director was appointed to represent all of the science *in hominem*. The central management of the

Museum was now, as it still is, a small group, representing the main customer areas of the organization, with science as just one of them. The effect is that the science has been marginalized. The head of science appointed by Neil Chalmers was a man called John Peake, whom most of the scientists in the Natural History Museum regarded as a loose cannon, at least on one of his good days. He had charm, and a kind of undisciplined enthusiasm for new things, rather like Mr. Toad in *The Wind in the Willows.* He also had an extraordinary effect on suits, which would become hopelessly rumpled even if they had been bought from Austin Reed just an hour beforehand. With wild hair and a lopsided grin produced by many years of smoking a disgusting old pipe, he often looked like the kind of geezer who approaches you on stations to request a small loan. I found it rather hard to dislike him.

Nonetheless, to have John Peake in charge of the taxonomic mission was analogous to having Count Dracula in charge of a blood bank. To be fair to him, he did some good things, such as directing the Museum towards molecular work, and equipping us with electronic mail long before most other institutions. He was doubtless forward looking. But the loss of whole areas of research was in my view a mistake from which it will be impossible to recover. Birds are the most popular of all animal groups among the lay public, but research in them was downgraded. It is a measure of the importance of this group of animals to the amateur naturalist that the Royal Society for the Protection of Birds spent £26 million in 2005–6, and it is at the forefront of habitat conservation, a cause with which any great museum should surely be identified. The bird collections—huge numbers of mounted specimens, skins and eggs—have since 1974 been held at the museum at Tring, Hertfordshire, which was bequeathed to the larger institution at South Kensington by Lord Rothschild in 1938. Mammals were historically one of the most important areas of study, but that has been reduced to "care and maintenance" of the collections. The argument was used that virtually all mammals and birds have been discovered, even though, of course, that leaves everything else about them still to study. My friend the spider man Fred Wanless was taken away from the animals he loved and told to work on nematodes, although there are plenty of spiders still to discover. The same story was repeated many

A sketch of John Peake made on a napkin at a conference

times around the Museum, and it is unnecessary to produce a litany. The Trustees had to approve all this. No doubt they had in mind the business model whereby science would eventually become self-funding from external grants, and less dependent on a dwindling Grant-in-aid from government coffers. The subject areas must be shifted towards those that would most likely secure their own funding. I doubt whether David Reid's great *Littorina* monograph would have been possible under this regime.

The Trustees are not, however, immune from attack. In 1960 the fish man, Denys W. Tucker, was dismissed from his post. In the 1950s Tucker had been noted for his research on eels. He was one of the protagonists in the discovery that these remarkable animals breed in the Sargasso Sea. Their mysterious disappearance from marshes into the sea, and their reappearance as baby eels, or elvers, in the same rivers and creeks, was a biological conundrum that needed solving. However, Tucker also became involved with the infamous Loch Ness monster, which, along with the Abominable Snowman, is the emblematic animal of cryptozoologists.* There is little doubt that the Museum establishment of the time looked askance at any endorsement by a member of staff of pseudo-science. Tucker persisted in his belief that in the deep waters of the Scottish loch lurked some kind of large vertebrate, often assumed to be a plesiosaur surviving from the Cretaceous. Warnings from the Keeper to keep away from "Nessie" went unheeded. But Tucker was a member of what his contemporaries would have described as "the awkward squad." The more he was instructed *not* to do something, the more he felt he must stick to his guns. The "Nessie" enthusiasts did nothing to help his case, naturally enough citing an authority from the Natural History Museum as a token of their own respectability. Tucker himself had fired off a series of memoranda and complaints about the incompetence of management "upstairs."

Eventually, disciplinary proceedings against Tucker were started, and, once initiated, the machine ground inexorably on. On 18 May 1960 the Director of the time, Sir Terence Morrison Scott, sent him a blistering memo proposing his dismissal "on the grounds of your long continued vexatious or insubordinate or generally offensive conduct." Finally, the matter reached the level of the Trustees, and their decision resulted in his removal from the staff. Being the obstinate man he was, Tucker fought back, asserting wrongful dismissal, and questioning whether the Trustees had the right to dismiss him. The matter even went to Parliament. The odd thing was that the official list of charges did not mention "Nessie"—rather, he was accused of the heinous crime of not filling in his diary. Possibly the Loch Ness monster was just too embarrassing and difficult, so his persecutors needed objective technicalities. The case ground on and on. In the end he sued the Chairman of the Trustees, who happened to be the Archibishop of Canterbury, Lord Fisher of Lambeth. We have seen already that Trustees at that time were deemed beyond reproach, and this particular Trustee obviously came with an impossibly high endorsement.

In the end, it required a ruling of the Court of Appeal to terminate Denys Tucker's campaign. It had taken many years of his life: "Eel

*Cryptozoologists are devotees of the theory that large unknown animals still lurk in remote corners of the world. Large lakes are favoured habitats; the Himalaya and Amazon and Congo rainforests are also popular.

A nineteenth-century view of Loch Ness, home of "Nessie"

expert loses 7 year battle" was the story in the *Daily Mail* on 8 December 1967. From this distance it does look as if his treatment was harsh, but even his friends recognized that his intransigence did not help his case. The saddest aspect was that Tucker was excluded from all parts of the Museum not accessible to the general public, which included the libraries. His career was castrated. I should finish the account of this affair by noting that modern surveys of underwater Loch Ness using the latest sonar equipment have failed to reveal a pod of "Nessies," although in the intervening years since the Tucker affair there has been no let-up in reported sightings. The climax of these "observations" came in 1975 when the ornithological artist and conservationist Sir Peter Scott gave "Nessie" a published scientific name (in the journal *Nature,* no less)—based upon a new photograph only slightly less blurry than its predecessors. That name was *Nessiteras rhombopteryx;* the species epithet referred to the allegedly rhomb-shaped limb of the Scottish "monster." It was quite quickly observed that the name was also an anagram of "monster hoax by Sir Peter S."

. . .

It is hard to believe that it has anything to do with the fishes they study, but the ichthyologists at the Natural History Museum tended to be a peculiar, if interesting, lot. Peter J. P. Whitehead was an expert on that most economically important group of fishes, the clupeids—herrings, anchovies and their numerous relatives. Many of those fishes with apparently endless, tiny forked bones are clupeids. The bones get stuck in your teeth and snag in your throat, but the fishes feed the millions. They are also full of Omega-3 oils, which make them good for the heart as well as the appetite. Just at the moment, they are considered to be almost the ultimate health food: if we ate nothing but lettuce and herrings we should all last till we are 103 years old. Whitehead's great work, published by the Food and Agriculture Organization (FAO) in 1985–88, is *Clupeoid Fishes of the World;* one part of it alone is 579 pages long, and of course it will be completely unknown to the average reader. Nonetheless, it is an heroic monument to a lifetime of study at the important end of biodiversity, where to name and identify are also to help people survive. Peter Whitehead followed, and possibly exceeded, Tucker's example of egocentrism. He was extraordinarily arrogant, but also good at what he did. Thin as a lath, with sharply defined bags beneath his eyes, he was perhaps rather an unlikely Lothario, but he was famous for his affairs with younger female members of the Museum staff. Apparently, he was irresistible. He lived somewhere beyond Oxford, and when he commuted into the Museum he travelled in the First Class end of the train, where in those days one could even have breakfast. He always claimed to be able to fix up a date with any attractive waitress by the time he reached Paddington Station. He came from a rather aristocratic background. His twin brother, Rowland, is a baronet: an urbane and cultivated man, with a successful City career and a number of charity commitments. When it came to inheriting the title, Peter evidently believed that he should be the "Sir" rather than his brother. I understand that this rare case might be determined by precedence—the first one out of the birth canal inherits the title. Peter took his brother to court over his right to the title, and eventually lost. Nobody could prove that he was the lead arrival. Not surprisingly, this

Peter Whitehead at home among the fish collections

verdict resulted in estrangement between the twins. Rowland recalls receiving letters of astounding virulence. Herrings evidently do nothing to suppress bile.

It is hardly surprising that the reorganization of science within the Museum in 1990 was anathema to Peter Whitehead. He foresaw the destruction of what he considered its traditional values of scholarly research pursued without too much red tape and leading in its own good time to the production of the definitive work. To such an individualist, the whole notion of being "managed" was preposterous, and he was angry about the arrival of what one might term Corporate Man at the Museum. Neanderthal Man was fine in the place, but Corporate Man belonged elsewhere. He was a broadly cultured man, with a passionate interest in classical music. I think now he would be held up as an admirable example of one who ignored the arbitrary division between arts and sciences. He wrote a satirical roman à clef under the title "The

Keepers" as a kind of warning, an exaggerated, farcical version of what he saw as happening around him. Sadly, the novel was never completed. The unfinished version has been circulating around the science departments in a kind of samizdat ever since Peter's death in 1992. The synopsis of the novel has every Keeper in love with some woman or other around the Museum. Wives are not mentioned, which probably reflects Peter's priorities rather well. The convoluted plot involves Brightlook Investments sponsoring various "improvements" that finish up with the gardens converted into a multi-storey car park and the galleries into an expensive massage parlour. Bluebottle Chemicals pay for what is entirely practical research: farewell trilobites and butterflies, hello germ warfare. After several botched assassination attempts upon him, the Director responsible for the whole débâcle is impaled on a rhinoceros horn. You get the general drift. I wish that the writing had been as good as the outline. I think it would have benefited from more distance; Peter was too involved with the thought that "the majestic grandeur of the Museum, its very dignity, was gradually eaten away, gnawed from the inside until the fabric was hollow, like a street façade from a film set merely propped up on waste ground . . ." Well, we *were* actually approached for a McDonald's franchise in the eastern basement.

Peter Whitehead became famous briefly in the wider world of non-ichthyologists for discovering a lost Mozart manuscript. It is surprising where the pursuit of herrings can lead. He was searching for a sixteenth-century work that included very early illustrations of Brazilian herring—"the *Libri Picturati* containing some of the most celebrated natural history drawings," as he described it himself. The same collection of manuscripts included a good deal of manuscript music by composers both illustrious and obscure. His detective work was persistent—the file of correspondence relating to it runs to two thick folders in the Museum archives. The original collection was removed from Berlin during the war, and he deduced that it "went to Kloster Grüssau, a Benedictine monastery in Silesia, almost certainly survived the war, but now [is] officially lost." After much dogged research, he eventually traced the collection to the Jagiellon Library in Kraków, Poland, where much of its natural history and music was safely preserved, including a Mozart score. Fortunately for Peter, his discovery was corroborated;

Dr Whitehead discovers something fishy

THE SUNDAY TIMES, APRIL 3 1977

THE GREAT MUSIC FIND

NIGEL LEWIS describes the hunt for lost manuscripts of Mozart and Beethoven masterpieces

Extending beyond the fishes: Peter Whitehead's musical discovery makes the headlines.

had it proved bogus, the newspapers would have had a great time with headlines featuring Red Herrings. The *Observer* and *Sunday Times* published the full story on 3 April 1977. When he was diagnosed with a brain tumour in 1991, Peter simply disappeared. He went off to Mexico. I am told that his wife visited the Museum shortly afterwards and tried to get into his office, but the Keeper of Zoology had locked it. It was too full of incriminating evidence concerning his exciting love life. The last we heard of him was a postcard sent to the fish section; it showed Peter

Whitehead on a small boat somewhere sunny in the company of a topless Mexican beauty. However, he did return home to die. It was a sad end.

Humphrey Greenwood, who was a near contemporary of Peter Whitehead's, worked on the cichlid fishes of the African great lakes. I first met him in a part of the basement known as Lavatory Lane. No single locality better encapsulated the class system of the old Museum. There were two different lavatories; one was labelled "Scientific Officers," the other labelled "Gentlemen," where the other ranks were allowed to pee. However, in both of these establishments the lavatory paper was a remarkably hard, shiny variety that had the words "*Government Property*" printed on every sheet. Presumably you were meant to reflect on your Civil Service grade even as you wiped your bottom. Humphrey and I exchanged humorous comments on these curious arrangements. After that, I would see him in the alley or colonnade at the back of the Museum where the smokers congregated by the bike sheds. Humphrey was a furious smoker who often paced up and down as if trying to undo the damage with a brisk walk. I learned about his fish work while puffing in a more leisurely fashion on a Gauloise. The cichlid fishes are a natural evolutionary experiment. They have evolved in isolation in localities such as Lake Victoria and Lake Malawi to produce dozens of endemic species. Humphrey made his reputation from the study of this "species flock," on which he published the definitive work, showing how behaviour differences and specialized feeding and breeding habits allowed all these closely related fishes to coexist. For example, there are species that feed only on snails and have special mouth parts to help them do this. One species became "mouth brooders"—the female carried the young inside her mouth, thus increasing the chances of the tiddlers growing to adulthood. Even more remarkably, there is evidence that Lake Victoria dried out twelve thousand years ago, so all this evolution must have happened rapidly since that time. This cichlid fish story tells us much about the generation of biodiversity, and how animals exploit their habitat, even though it may not be typical of animals in general. Furthermore, what has happened subsequently in Lake Victoria certainly tells us about what damages ecology when an alien species is introduced. The Nile perch (*Lates niloticus*) was introduced into the lake as a commercial fish in the late 1950s, because it grows rapidly to a great size. It is a living example of fast food. The effect was to drive some cichlid species to the brink of extinction. Humphrey said wryly that he sometimes did not know whether he was godfather or obituarist to some of the fishes he loved.

Humphrey's cichlid work is a good example of how research spans generations of Museum employees. The tradition of working on the fishes of the African lakes goes far back in our scientific history. In 1864 Albert Carl Ludwig Gotthilf Günther, who was Keeper of Zoology for twenty years, had described a number of species from dried skins collected from Lake Malawi on the second David Livingstone Zambesi Expedition. The zoologist Georges Albert Boulenger then produced a monumental *Catalogue of the Freshwater Fishes of Africa in the British Museum* between 1909 and 1916, of which the cichlids formed an important part of the 1915 volume of that work. The small matter of the First World War did not dent his endeavours. I should add that Boulenger's work in general is still cited regularly in the scientific literature; proof, if you like, that taxonomic contributions can have durability that more flashy publications lack. Charles Tate Regan, who was Director of the Museum between 1927 and 1938, produced a work on Lake Malawi fishes in 1922. He was followed by the redoubtable Ethelwynn Trewavas, the last of Humphrey's predecessors, and she, too, made her contribution to cichlids from the 1930s onwards, especially those of Lake Malawi—sometimes in combination with Regan. Ethelwynn worked on many other kinds of fishes in addition, including eels and strange, ugly-looking deep-sea species.

The collections of all these workers survive for future generations of scientists to study, so that an interested visitor will feel a kind of tactile link with his intellectual forebears. If he opens one of the appropriate jars where the cichlids are preserved he can touch a specimen that was handled by Trewavas, that had once flipped and jumped in its death throes before being preserved in spirits. A vision of an expedition long ago comes into his mind: Ethelwynn in her sensible khaki clothes deep in Africa, issuing orders to porters, busy with her jars. I had first come across these fishes lurking in the dark in the old Spirit Building when I explored the nooks and crannies of Gormenghast. Now they live in bright new storage in the glass palace of the new Darwin Centre at the west end of the Museum site, rank upon rank in transparent jars. They occupy their appropriate place in the twenty-seven kilometres of shelving where the "wet" collections are stored: lizards on one floor, octopuses on another, snakes elsewhere again. In the basement of the Darwin Centre there is a battery of tanks that house the giants of the fish collection. The artist Damien Hirst leapt to fame with his pickled shark: well, here are a dozen unsung Hirsts wallowing in their formalin archive. It is a development to be welcomed that the public are now allowed to tour through these historical collections. The current fish man, Darrell Siebert, told me that he had found a species in Borneo that was a staple food for a whole village but had never even been named and described—and this at a time when deforestation was changing its habitat for ever. Surely it is not just what E. O. Wilson termed "biophilia" to believe that for a whole *species* to die out before it has even been named is a tragedy commensurate with any of those that human beings have devised for each other.

Ethelwynn Trewavas was devoted to Denys Tucker; indeed, they may have been lovers at one time. When Ethelwynn made her retirement speech, she let the management have it from both barrels about what she regarded as his shabby treatment. But she continued to work on fishes until she was very old, sprightly to the last. Humphrey Greenwood went on from strength to strength as he documented the cichlid evolutionary radiation in Lake Victoria. His 1974 book *The Cichlid Fishes of Lake Victoria: Biology of a Species Flock* is a classic of biology. He was elected to the Fellowship of the Royal Society, which is one of the very few clubs for which membership cannot be bought or traded. For a while he was the only FRS in the Natural History Museum. Nonetheless, he could not be described as a happy man, any more than Peter Whitehead or Denys Tucker. During his smoky marching up and down the colonnade he grumbled and moaned about the iniquities of management. He frequently had dark rings under his eyes, speaking of disturbed rest and late nights pacing the carpet. Nor did he enjoy easy relationships with his students; he was so hypercritical, they said. He was evidently a difficult and temperamental man. His death, too, had more than a whiff of tragedy. He died of a heart attack while working late in his laboratory. Had a night warder discovered him he might have been saved. So the curation staff in the 1980s who had to work for these awkward scientists often gnashed their teeth in frustration, although frankly they, too, included a number of eccentrics. Jim Chambers was a tall, lugubriously humorous man with wild hair running all down his back. He wore John Lennon glasses, and played in a group vaguely reminiscent of the Bonzo Dog Doo-Dah Band. His barking laugh could be heard all around the Museum. Jim left voluntarily in the aftermath of the 1990 "night of the long dissecting knives."

I hesitate to add yet one more fishy figure, but the story would not be complete without introducing Alwyne Wheeler into the plot. His finest memorial is probably his contribution to cleaning up the River Thames. London's river was a disgrace after the Second World War—increasing and unbridled pollution and waste had removed oxygen from its lower stretches, until it had a distinct pong in summer, which I can remember sniffing queasily as a child on visits to the Embankment. About the only creature to thrive in the fetid mud was the little red *Tubifex* worm, which is almost impossible to kill. My father used to dig clumps of it up to use as fish food. Other aquatic life, and especially piscatorial life, was severely on the wane. "Wyn" Wheeler was much more the populist than either Whitehead or Greenwood. He helped set up networks such as the Thames Estuary Partnership, bringing the interests of anglers and industry and government agencies together. Nowadays there is a host of such organizations ranging from the London Biodiversity Partnership to the Thames Salmon Trust, most of them seeking co-operation between partners with an interest in a clean river. Wheeler worked tirelessly identifying fishes brought in by members of the public. Sometimes there were surprises, as when "tropical" fish were found living near the warm-water outlets of power stations. He wrote popular guidebooks. He enjoyed friendly relations with watermen up and down the river.

The last naturally spawned salmon was caught in the River Thames in 1833. One hundred and forty-six years later the Thames Salmon Rehabilitation scheme attempted to reverse history: fish were introduced artificially. Such was the progress in cleaning up the former "open sewer" that by 1982 a spawning salmon run up the Thames happened once again. There was a tremendous sense of triumph when this was announced, for salmon are choosy fish—they will not tolerate pollution. Now more than twenty species of fish can be found in the Thames. And with the fish comes all the associated bird life: richness has been restored. "Wyn" Wheeler was one of the heroes of this triumph of conservation and habitat restoration, which remains a model for many others around the world. It is not a battle that has been won, however. Exceptional rains in 2005 strained the antiquated sewerage system of London beyond breaking point and raw waste ran into the river once again, which shortly yielded up a slurry of scaly corpses. The price of freedom from pollution is eternal vigilance.

Wheeler could not bear Peter Whitehead. He hated his aristocratic mien, and regarded his ostentatious cultural breadth as not a little precious and pretentious. Whitehead fully reciprocated the loathing. Fortunately for him, he was the senior employee at a time when seniority counted. The two fish men never talked to one another, if they could possibly avoid it. The upper echelons of the Museum regarded the *whole* of the fish section as almost unmanageable. In Wheeler's Annual Confidential Report in 1979, the Keeper described the "problems . . . inherent in the fish section manned by a number of 'prima donnas.' " It may have been a response to this perceived management impossibility, but by a dramatic coup de théâtre that could have happened only in the fish group, John Peake decided to reverse the hierarchical structure: he put Alwyne Wheeler in charge of Peter Whitehead. This was a ludicrous thing to do, because even "Wyn" Wheeler's supporters would have said that he had little talent for managing anything, even his temper. He soon obtained satisfaction for years of slights by vetoing Peter Whitehead's publications on musical or cultural matters—after all, he would argue, the job has to do with fish, not all this other stuff. It was either hilarious or tragic depending on where you stood: through it all the curators struggled on with their duties. When I talk about those times to the surviving curators, such as Ollie Crimmen, they are recalled with a mixture of affection and exasperation. About "Wyn" Wheeler Ollie said: "He taught me everything I know; he drove me mad."

I have dwelt on the drama within this group of fish zoologists not only because it makes entertaining gossip, but because it illuminates what Museum science is probably all about. I hazard a guess that if any quality assessment of fish research had been made during the period 1970–90 the Natural History Museum would have come out as one of the world leaders, possibly *the* world leader. This was while feuding, "primadonnaing," *affaires de coeur,* litigation and heaven knows what else were happening in all directions. Management was concerned, as it always is, with what Sir Walter Bodmer called the "right to manage." This does not have any necessary connection with what actually happens in the scientific arena. It might actually be the case that having an obstructive, disagreeable, temperamental, competitive or even downright anarchic group of people all scrapping and competing is the best way to push knowledge forwards. Should Peter Whitehead have turned his back on important musical discoveries just because it said "fish man" on his label? I usually avoid such rhetorical questions, because they permit of only one answer, but in this case I'll go ahead: "Of course not!" Research is sometimes a matter of sheer serendipity, but if anything it is more a matter of persistence, of sheer bloodymindedness. The kinds of people who are able to work alone for long periods, often without much encouragement, with modest financial reward, and by remote African lakes with no creature comforts, are actually rather rare. These are ideal employees of a great museum.

Sometimes individualism merges into lunacy. One of the most striking examples in the zoology part of the Natural History Museum was the case of the sponge and protozoan worker Randolph Kirkpatrick. In William T. Stearn's official history of the Museum he merits a peripheral mention as an author of *The History of the Collections* (1906). He actually did some quite respectable work on sponges. But he is known today for one of the more extraordinary intellectual excursions in the history of palaeontology. In the early Tertiary rocks a rather common kind of fossil is the nummulite (see colour plate 8). It is about the size and shape of a coin, and is made, as are so many fossils, of calcium carbonate. It is a fossil of a gigantic single-celled organism, a behemoth of the group known as Foraminifera—which are usually not much larger than a pin head—and for which Kirkpatrick had responsibility in the Museum. Inside the "coins" are a beautiful spiral series of tiny chambers that were secreted by the protoplasm of the simple organism. Limestone rocks containing nummulites polished by the action of the sea can look like nebular maps of some remote corner of the universe. Nummulites are sometimes so abundant as to make cliffs and whole rock formations; they are most celebrated as supplying the nummulitic building stone for the pyramids of Gizeh in Egypt—hence *Nummulites gizehensis.* I have seen them peeping out of pebbles on beaches in Greece. They are certainly not rare.

But Randolph Kirkpatrick managed to convince himself not only that they were common, but that they were truly ubiquitous. He believed that *all* rocks were made of nummulites. The case for sedimentary rocks was at least conceivable because these kinds of rocks are mostly laid down under water, and nummulites lived in shallow seas. But when Kirkpatrick's ideas took their curious hold on him, he managed to convince himself that even igneous rocks, the kind that are erupted from volcanoes, were made of nummulites. He privately published his ideas in a series of books under the title *The Nummulosphere.* Thin sections of volcanic rocks like basalt are shown with the nummulites "dotted in." They remind me of those star maps where Scorpio or Aries are drawn in from among the swath of stars in the sky. All the stars that don't fit are ignored. Since basalt rocks include patterns of tiny swirling felspar crystals, it is possible to "see" nummulites by ignoring all the little specks in between. I own a precious copy of Part III of *The Nummulosphere* published in 1906. These outré ideas are possibly the end point of the freedom to pursue a line of enquiry unchecked. Stephen Jay Gould wrote in *The Panda's Thumb:* "I respect Kirkpatrick both for his sponges and for his numinous nummulosphere. It is easy to dismiss a crazy theory with laughter that debars any attempt to understand a man's motivation—and the nummulosphere is a crazy theory. I find that few men of imagination are not worth my attention." I agree that a few nummulospheres are probably the price that we should all be prepared to pay to permit intellectual freedom and joyfulness. After all, some ideas initially thought to be crazy turn out to set the course to the future, and their originators then know a special kind of satisfaction. As Francis Bacon put it in one of his essays in 1625: "no pleasure is comparable to the standing upon the vantage-ground of truth." As we have seen, there are also delusions about the truth, but one wonders whether those that hold them still feel Bacon's incomparable pleasure.

Not far from where Humphrey Greenwood used to smoke furiously

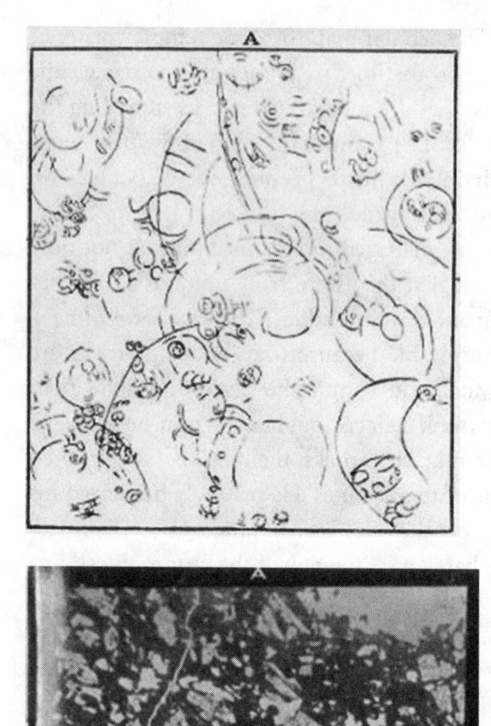

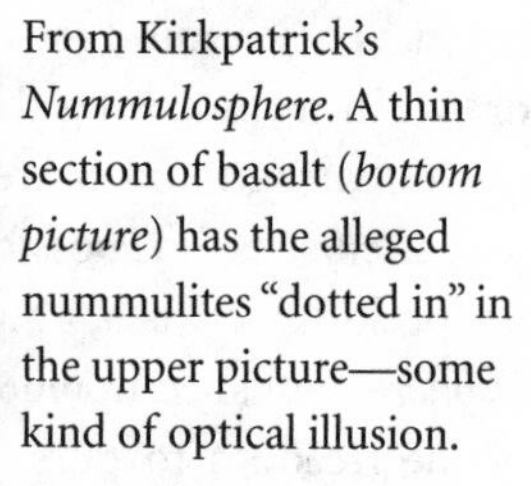

From Kirkpatrick's *Nummulosphere.* A thin section of basalt (*bottom picture*) has the alleged nummulites "dotted in" in the upper picture—some kind of optical illusion.

out at the back* of the old building, there is a pair of modern glass doors, which indicate the entrance to the Molecular Biology Laboratory. Passing through these doors from the old collections is an odd

*As this is written (2007), beyond this point at the west end of the building there is nothing—an open space where the old Spirit Building once stood. This is where the second stage of the Darwin Centre will appear.

experience, and the first-time visitor might briefly wonder whether he is in a natural history museum at all. The whole place exudes a cleanliness that even the best-maintained collection of ammonites never quite manages. White lab coats must be worn at all times. The visitor is subliminally aware of the hum of machines. Peering in through windows into the laboratories he spies legions of chemicals in bottles, and pipettes, and fume cupboards, and laptops: in fact, it is just what the man in the street might think a laboratory should look like. This is where the molecular sequencing work is carried out, and I have already given sufficient examples to show how this work now melds in perfectly with the more traditional research of the Museum on the taxonomy of snails or primitive plants. The old collections and this buzzing, busy laboratory are bedfellows after all. The laboratory has grown and changed as the technology has improved. This is a common story in science: what starts as hand-cranked and time-consuming soon becomes automated and routine. The quantity of tissue material required for DNA analysis has been progressively reduced, most importantly following the development of the PCR technique that "multiplies" even the tiniest gene sample.

Through the several rooms bustles Andy Warlow; he is small, very Welsh, indefatigably cheerful and bursting with proprietorial pride. Like all scientists with a technical bent he refers to machines as "kit," as in the phrase "this is a nice piece of kit." It is a fact that most pieces of kit are more or less rectangular boxes with dials on. It is what goes on in the innards that varies enormously. Many pieces of kit are also expensive. The automated sequencer produces the kind of molecular data that help with taxonomy and systematics and costs a quarter of a million pounds; it is served by a robot which now automatically does the kind of work that formerly had to be done by laboratory technicians, such as purifying samples and preparing them for analysis. It has little automated arms like a Dalek. These pieces of kit came into operation in 2003 and 2004, respectively, so they are what the scientific cliché always refers to as "state of the art." In the modern Museum their services must be purchased by "customers," that is, scientists, who are charged £2 a sample in-house; any contract from outside will pay four times that sum.

The technician in charge of the machines tells me that there are something like seventy research projects currently using the facility. One I happen to know about is Dr. Tim Littlewood using sequence data to find out more about the relationships of the mysterious sea spiders, or pycnogonids, which include some of the oddest-looking of the arthropods—all legs and no body. It so happens that Derek Siveter of Oxford University has described the most perfectly preserved fossil of a sea spider from the Silurian rocks, some 425 million years old—it looks for all the world like one still living. Why are they so unchanging? Are they related to scorpions—and the true spiders—or are they simply very odd crustaceans? The molecules will hold the key to answering these questions, for they carry the narrative of common ancestry locked away in the genome. The machines are there to help crack the problem.

In another room Andy Warlow shows off some giant refrigerators. These include machines storing the frozen-tissue collection: an archive of flesh, the soft parts of organisms that normally decay. At minus 85 degrees centigrade flesh retains its genetic information indefinitely. Hence tissues that once formed the basis of molecular sequencing experiments can be stored here for future investigators. This archive, suspended from the processes of decay, also saves unnecessary culling of wild animals to secure further samples. In the old spirit collections flesh is preserved in the pallid ranks of fish and prawns sitting in their jars. Sadly, the employment of smelly formalin as a preservative also served to destroy the DNA by unhitching all the chemical bonds that hold it together. The curators were unwittingly tearing up the record of the genes even as they sought to defy the effects of time. Ironically, the oldest "wet" collections of all are still often useful as the source of DNA samples. This is because these specimens really *were* preserved in spirit—in alcohol, no less, and alcohol does not destroy genetic information. Genes, like old soaks, can evidently be pickled in spirits. There are stories of specimens being bottled in gin or brandy when regular supplies of preserving alcohol had run out, which says admirable things about the priorities of the collectors. Dried skins, and even bones, also preserve significant amounts of DNA. The specimens stored in formalin jars are still useful, of course, but only for looking at morphology. On the other hand, neither will the frozen-tissue collection ever replace the whole specimen, because it only tells part of the story. The same cold room houses another freezer that holds the Museum's part of the Frozen Ark project, rather unfortunately known by the acronym FARC. This initiative is designed to preserve the genetic information of threatened species around the world. Perhaps the idea of resuscitating such species using the frozen genes is beyond current capabilities, but the urgency of having such an archive is obvious already, and will be more pressing in the next century. Andy Warlow tells me that within the freezer lies a sample of *Leptodactylus fallax*, known as the Caribbean mountain chicken, and now in imminent danger of extinction in its native Montserrat and Dominica, where it lives above 300 metres and feeds on crickets. It is a large frog. But its common name betrays the reason why it is so nearly extinct.

Another laboratory is dedicated to working on the schistosome parasite. The disease caused by this particularly unpleasant organism, schistosomiasis, is perhaps better known as bilharzia. More than three hundred million people are infected by this debilitating condition, so this is one case where I do not have to explain the wider significance of Museum research. The name recalls the discoverer of the parasite, Theodore Maximilian Bilharz. Poor Bilharz died young in 1862 from another complaint altogether—typhus, caused by a malevolent microorganism. There has been a long tradition of studying bilharzia and its causes in the Museum. The man in charge when I joined the Museum was Chris Wright, a very effective scientist who was also definitely a toff in the tradition of old-style Museum employees, but with none of the superior airs of Peter Whitehead. Like Bilharz, he died tragically young, of cancer, in a way unrelated to parasites. Vaughan Southgate, who succeeded him, is still in vigorous health: he is bluff and florid, affable and clubbable, dedicated to game fishing, with a laugh like a steam train going uphill, and one feels that there is no malevolent parasite that he could not slough off with the aid of a couple of stiff gins. Nonetheless, he often returned from African fieldwork during the 1980s harbouring a parasite or two, which provided much entertainment for the experts at the London School of Hygiene and Tropical Medicine. On one occasion I asked him anxiously on his return what he had brought back with him this time. "No idea, old boy," he chortled amiably. "It hasn't hatched out yet." David Rollinson is currently head of the schistosome group; it used to be known as "Experimental taxonomy," because the biologists used their work as a testing ground for all manner of new molecular and computational techniques. The laboratory is currently a World Health Organization reference centre for the identification of schistosomes and their snail hosts. I always hoped that the head of this laboratory would be known as Doctor WHO.

I should explain the life cycle of the parasite at this point. *Schistosoma* is a specialized kind of flatworm that lives in the bloodstream in humans, and many other species can be infected. Unlike any other parasite it has separate sexes. Adults have been known to live for up to forty-three years; the body is about a centimetre long. Once paired with a male, the female *Schistosoma* lays thousands upon thousands of minute eggs. These are what cause the pathological symptoms, for they fetch up in the liver. The immune system "wraps them up" in what is known as the granuloma immune response. Eventually this causes severe organ damage, sufficient to cause jaundice and liver failure. In children, distended bellies provide a characteristic sign of infection. There are nineteen or twenty *Schistosoma* species, of which three commonly infect humans: the *S. mansoni* species group; the *S. haematobium* group; and *S. japonicum*. As the name suggests, the last is characteristic of the Far East, but was eradicated from Japan in the 1940s, persisting today in Cambodia and Laos. Bilharzia poses the biggest health problem in the wetter parts of Africa, although the disease was also carried to the Caribbean as another unwelcome consequence of the slave trade. The parasite burrows out through the gut or bladder wall to distribute its eggs—blood in the urine may well indicate its presence. Once in the water, the larval schistosome seeks out its next host, a freshwater snail. Inside the host's mantle it completes the next stage of its life cycle. Then it swims out as a "fluke" again to find a human victim, boring in through the skin to enter the bloodstream, where it can stay long enough to ensure the continuation of the lethal cycle. Biologically speaking, it is an elegant way to feed off other organisms while ensuring the survival of the parasitic genes—or "shellfish genes"!

The young schistosome man, David Johnson, who takes me around the laboratory, has an undertone of admiration when he talks about these cunning organisms. The larvae are quite specific about the snail hosts they will infect: the *S. mansoni* species group goes for the snail *Biomphalina,* while the *S. haematobium* group will take up residence in another snail called *Bulinus,* of which there are about forty species. Interestingly, some *Bulinus* species seem able to "resist" infection by schistosomes, and some current research is trying to find out how. To eliminate the disease, people could eliminate the snail, or stay out of the water, neither of which is a practical solution in many poor countries. It is certainly a good idea to find out as much as possible about the foe. Many human parasites arose in Africa alongside our own emerging species, whence they spread around the world following the human diaspora. When the molecular evidence for the relationships of bilharzia parasites came out in 2000, the surprise was that the Asian species seemed to be the most primitive, rather than the African species. Just to prove that this links in with the systematic purpose of the Museum, this discovery also resulted in the "sinking" of an exceedingly inelegant generic name, *Orientobilharzia,* which had been given to the Japanese species. It is now considered that human colonization by schistosomes happened independently in the three main infecting species. Current research is focussing on the molecular identification of snails and the parasites they contain—each snail is like a sample bag containing its infecting schistosome, whose DNA can quite easily be extracted. The sequencing of the full genome of *S. mansoni* is in prospect, which may help with finding a novel target for a vaccine. The Museum is collaborating with a research station in Zanzibar, where there are apparently two species of *Bulina,* one of them being a resistant form. It seems that the distribution of the snails themselves is caused by water chemistry—and maybe this offers the prospect of a different kind of control. After all, if a resistant snail could be favoured, the parasite would be at a disadvantage. The questions continue, as they should.

Near the laboratory there is a living collection—tray after tray and tank after tank of water snails—more like a zoo than a museum. It is very hot, to keep these tropical species happy. The snails I see are planispiral—they are curved like a Cumberland sausage rather than spiralling upwards. My father called them "ramshorns." They are fed on nothing more exciting than lettuce. David Johnson tells me they are very fussy about their water, which must not have any heavy metals in it. There is a special Museum borehole to bring up pure water from the depths in plastic pipes to keep the thousands of little creatures happy. I cannot tell which ones are infected with schistosomes just by looking at them. But in this part of the laboratory infection can be experimented on under controlled conditions. This may one day lead to the relief of much human misery.

I have described the work of this part of the hidden Museum in some detail to show that taxonomic research does link directly with global issues of human health. This does not necessarily make the research more important, but it certainly does make it more fundable. The schistosome facility in the Natural History Museum secured the future of much of the molecular laboratory as a whole; it was set up using funding from the Wellcome and Wolfson Foundations, organizations that are interested particularly in health issues. The research staff is wholly admirable. If medieval saints demonstrated selfless compassion by washing the sores of lepers, then their modern equivalents could be those who attempt to improve the lives of thousands of people whom they have never met by wading through mosquito-infested swamps crawling with infected snails.

While offering a digression on saints, one is naturally reminded of sinners. Where the molecular laboratory now stands there was once a whale room; only the London brick walls of the old building remain, for inside all is now gleaming steel and machines. One of the Museum's great sinners used to hide there, Peter Purves, the whale man. Whales* were once an important part of the Museum's research programme. Whenever an interesting whale was beached on British shores it would find its way to the whale room. The origin of this rich source for the collections goes back to a statute of the time of Edward II, establishing that whales and porpoises are Fishes Royal, belonging to the Crown. By June 1912 this ancient right had mutated to become a directive from the Board of Trade to all Receivers of Wrecks that the Museum should be informed of strandings. The Museum prepared an identification guide

*Whales are part of the mammal group Cetacea, which also includes porpoises and dolphins; all of these organisms are equally the subject of scientific research and conservation efforts.

to help the men in the field. And the Trustees then asked the Treasury for permission to buy a pair of whaler's overalls—they ran a tight ship in those days when it came to expenditure of public money. I should say that the Directors of the Museum had an impeccable record when it came to alerting the authorities about the dangers of over-exploiting whales. Sir William Flower had delivered a speech in 1885 about the avaricious short-sightedness of the Atlantic and Australian whaling industry. Sir Sidney Harmer repeatedly drew the attention of the Trustees to the wholesale slaughter of the great whales in the Antarctic, including the largest animal of them all, the blue whale (*Balaenoptera musculus*). The model of this whale on display to the public in the Whale Hall was completed in 1938 and for a while was the largest in any museum. For commercially hunted animals the conservation argument is often not much more sophisticated than the warning in the fairy story about not killing the goose that laid the golden egg, and there is something very depressing about hearing arguments for continued exploitation today that would have been familiar to Flower and Harmer. Whales deserve to live simply because they are wonderful animals, and never mind the trade in ambergris. However, bulky stranded whales are not exactly easy specimens to deal with. You cannot pickle them, nor dry more than fragments. Anyone who has been downwind of a rotting whale carcass will know that, as they rot, they stink. Hence they were mostly preserved as skeletons. But when there was a new arrival in the whale house it imparted to the back arcade a very singular atmosphere. People did not linger. In the old days there was a "whale pit" at the western end of the building, where blubbery bodies could be buried to rot in their own time. Later, evil cauldrons of alkaline potions speeded up the whole business of decay.

One of the problems with whales is trying to work out how old an individual is when he or she is stranded. For effective conservation one needs to know how many years it takes a whale species to grow up, and at what age individuals can reproduce. It was discovered that whales lay down waxy layers in the inner ear, almost like the growth rings on trees. Peter Purves was extremely skilled at the delicate operation necessary to extract this information. A young whale man told me that in his opinion Purves did much of the work that helped a more famous Museum mammal scientist, F. C. Fraser, achieve election to the Royal Society in 1966. Nobody really knows the truth of this, but what is not in doubt is that Peter Purves was an inveterate drinker. As *The Sun* might put it, he was a "hell-raiser." In the 1970s he would weave his way down the front steps and out of the Museum at about the time that most of us were having our afternoon tea break. He had that rather delicate, cantilevered gait of the experienced toper. I could never work out the origin of his accent; it could have been Irish—but then again it could have been Scottish. He had the deliberate delivery of the habitually sozzled, a series of short barks separated by significant pauses. When he spoke, his sentences always made rather ponderous sense. But in this condition he could perform the delicate slicing necessary to age whales. Apparently he was much less adept while sober. It could be argued that alcohol was necessary to help him survive a life in the whale room with its overwhelming pong. It was reported that he had half-bottles of Bushmills tucked away in the blubber, but few people would have cared to poke around the rotting carcasses to find out. He was a short, dapper man, always rather well-dressed. He remained trim to the end. There was a Museum club, The Tetrapods, of which Peter was a member, which met periodically above a pub called the Goat in the evenings. After a vaguely relevant lecture on some aspect of natural history, nearly everybody got drunk, but in some strange way, Purves would stay on his feet long after other members were looking much the worse for wear; practice, I suppose. His boozing did land him in trouble on a number of occasions. Once he fell into the "stripping tank" in the whale room and suffered serious burns; he said afterwards that he would never have survived had he been sober. He was always in demand to show his dissection techniques. He was once invited by Professor Pillari in Berne to dissect the rare blind river dolphin. He fell to talking and drinking while on the leisurely railway journey through France. From the Gare de Lyon in Paris he knew that the train went straight to Geneva. Eventually he fell asleep in a stupor. When the train stopped, he woke with a start, assumed that he must have arrived at his destination, and woozily got off the train. He was actually at a border post, for the train had made an extra, unscheduled stop just outside Switzerland. The train moved on, leaving him standing on a small station in the middle of nowhere with

ABOVE The Natural History Museum in South Kensington.
BELOW The main hall of the Natural History Museum, showing the nineteenth-century pride in revealing the iron structure. The tail of *Diplodocus* is in the centre (*see p. 6*).

ABOVE Sir Richard Owen, inspirer of the Natural History Museum, splendidly painted by Holman Hunt.
BELOW One of the ceiling panels from the main hall, showing *Pinus sylvestris*.

ABOVE Time measured in tree rings—a section through a giant sequoia (*Sequoiadendron giganteum*) from the Sierra Nevada, California, on display outside the Herbarium.
RIGHT A fish preserved in a jar in the spirit collections; the moonfish *Mene maculata* in gloomy pose.
BELOW The truth of bones: a parade of large mammals in the osteology collections.

LEFT Discovering new species. Nathan Muchhala in the cloud forests of Ecuador holding the specialized flower *Centropogon nigricans,* which (*below*) feeds (and is fertilized by) the equally specialized bat *Anoura fistulata,* with its extraordinarily long tongue.

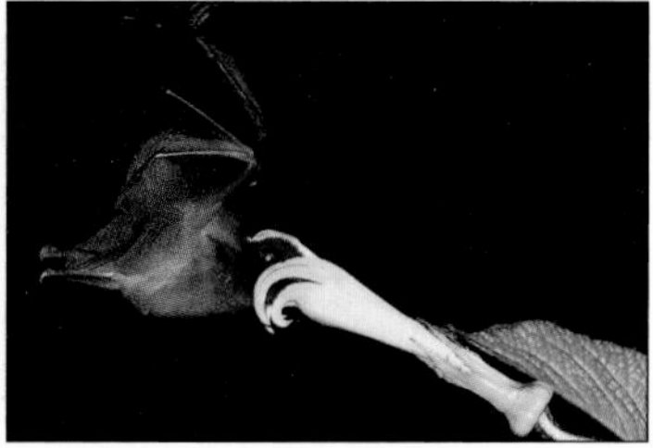

BELOW Modern collecting. New species of insect are frequently discovered by "fogging" tree canopies and collecting what falls out in special funnel traps.

ABOVE The acceptable legacy of empire: an ornamental fountain in the Botanic Garden in Christchurch, New Zealand.

ABOVE Carl von Linné (Linnaeus)—a portrait showing the great nomenclaturalist dressed in Lappish gear. From a painting by Martin Hoffman, 1737.

BELOW "Old Man Banksia" (*Banksia serrata*), the Australian shrub named in honour of Sir Joseph Banks from Cook's first voyage—the specimen is to the left and the illustration of it to the right. To be preserved in perpetuity in the Natural History Museum.

Behind the scenes in the Museum. Above, the leather-bound learned volumes and sports jacket of the traditional image; below, the white coat and dust-free conditions of the modern DNA laboratory.

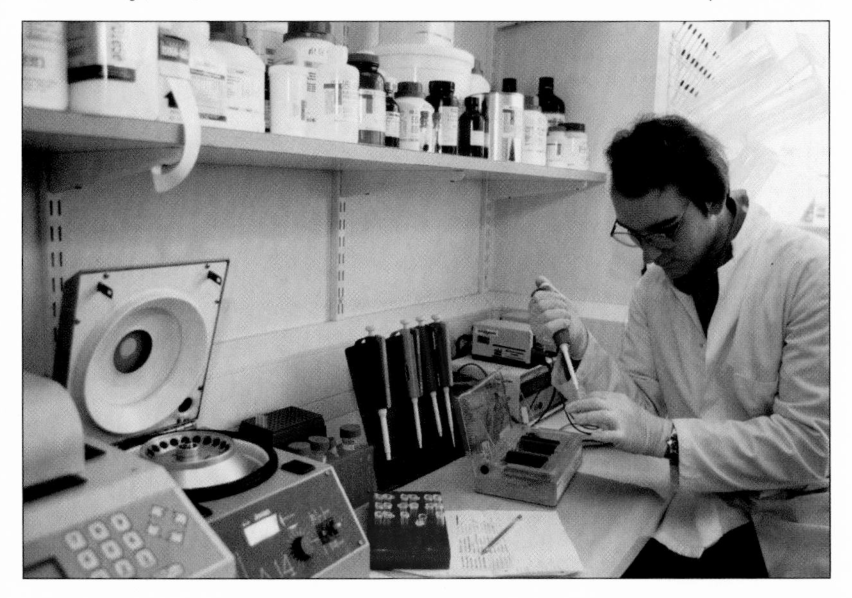

LEFT Living colonial bryozoan *Adeona* in the warm seas of South Australia (*see, for example, p. 298*).
BELOW Virtual image of the Silurian fossil ostracod *Colymbosathon ecplecticos* with its limbs and genitalia remarkably preserved. Below is its living relative.

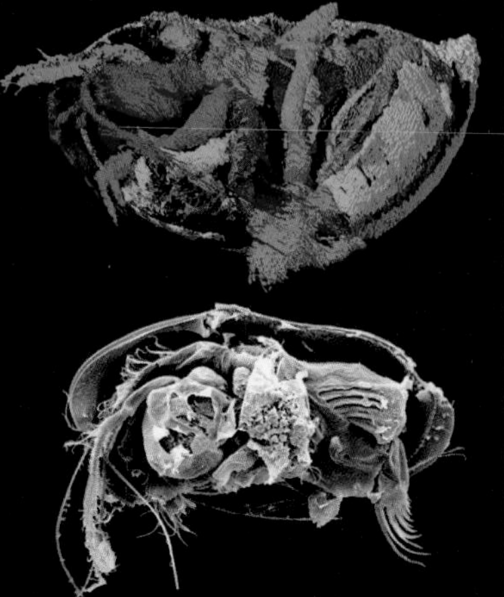

BELOW Entrepreneurial palaeontology: Devonian trilobites laid out for sale on brightly coloured fabric on the roadside in Morocco.

New data from old bones: braincase reconstruction (*right*) of the early bird *Archaeopteryx* prepared by Angela Milner and colleagues, with a full reconstruction to the left.

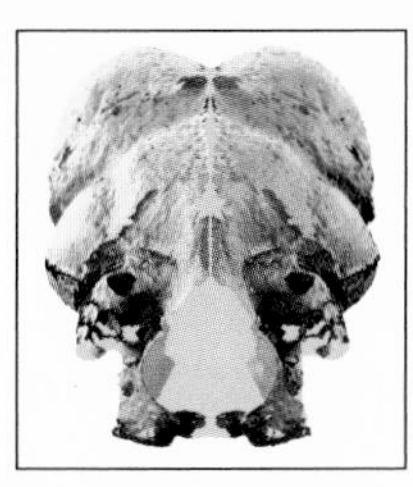

ABOVE Excavating the early history of mankind in Britain: digging at Boxgrove, Sussex.

LEFT Unconsolidated sediments on the Suffolk coast near these cliffs have recently yielded the most ancient evidence of humans in Britain (*see pp. 105–7*).

Cichlid fish have been a long-term subject of study among the zoologists: this is *Tropheus moorii,* one of many species endemic to Lake Tanganyika, Africa.

Coin-like fossils of nummulites from the early Cenozoic (Eocene), *Nummulites laevigatus* (Brugiere), Whitecliff Bay, Isle of Wight; see section through a fossil on lower right to understand what Kirkpatrick thought he saw in thin sections (*see p. 139*).

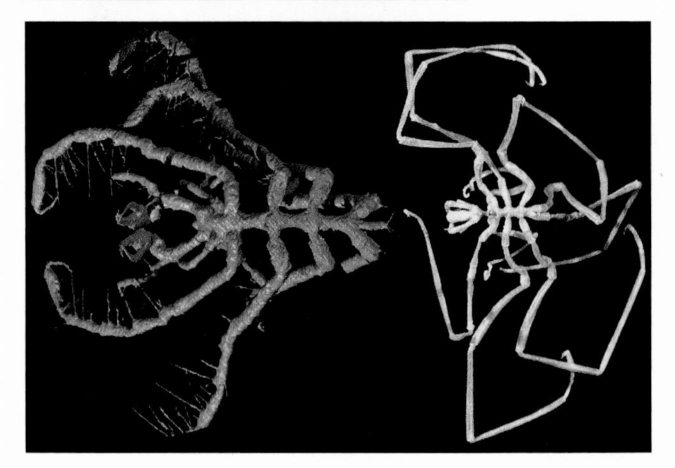

About as perfect as fossils can be: a sea spider (pycnogonid) reconstructed from thin sections through a hard nodule preserved with Silurian rocks in Herefordshire, England, by Derek Siveter and colleagues. At right, its living relative.

LEFT The commanding Sir Joseph Banks, Bt., the most powerful man in eighteenth-century British science (*see* Banksia *on plate 4*). Oil on canvas by Thomas Phillips, 1810.

BELOW Herbarium sheet carrying the cocoa plant, source of chocolate and happiness, *Theobroma cacao*, accompanied by its original drawing. This specimen was collected by Sir Hans Sloane in Jamaica, and forms part of the nucleus of the "BM" collections—still safely curated in the Sloane Herbarium in South Kensington.

LEFT Lichens growing on gravestones in Oxfordshire—the dates provide one way of computing the slow growth rate of lichens (*see p. 162*).

ABOVE One of the Linnean herbarium sheets curated in the Linnean Society of London, and forming the foundation of the naming of plants. The colour has faded a little, but the specimen otherwise survives well. This is the sweet pea *Lathyrus odoratus* L. 905.12 (LINN).

ABOVE A living plant that may be destined for preservation in the herbarium. This is a recently named relative of the tomato called *Solanum huaylasense* Peralta from Peru.

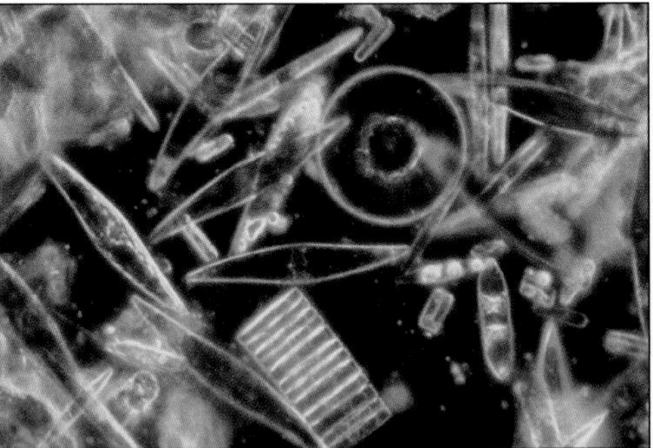

ABOVE Beauty at small size: the delicate silica tests of diatoms under the microscope (*see pp. 179–80*).

LEFT An attractive species of *Hypericum*, the genus that has been the lifetime's work of Dr. Norman Robson (*see p. 183*).

RIGHT "The Queen bee": Miriam Rothschild, doyenne of Museum trustees.

BELOW *Aleurocanthus woglumi*—a troublesome insect pest (*see p. 198*).
BELOW RIGHT *Encarsia perplexa*—a parasitic wasp that preys on *Aleurocanthus*.

BELOW The homemade field laboratory of Vane-Wright and friends on location in South-west Africa in the early 1970s.

LEFT A termite mound in outback Australia: one of the insect world's most sophisticated structures.

BELOW Now thought to be related to cockroaches: the primitive termite *Bifidtermes* at work on wood; note wingless larvae (*see p. 211*).

BELOW Laid out for study: an historically important collection of butterflies and moths made by Alfred Russel Wallace, Charles Darwin's co-author on evolution.

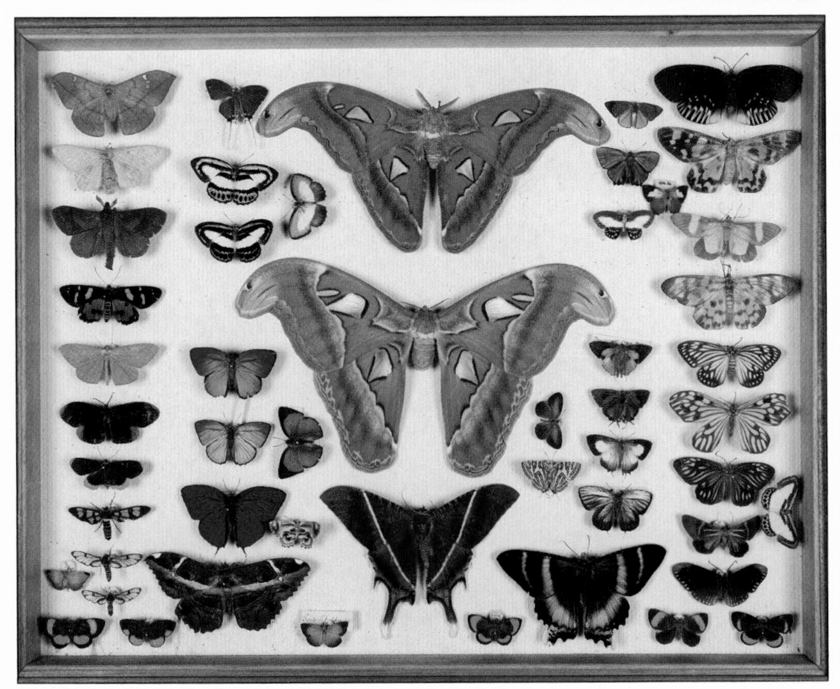

ABOVE The cutting of thin sections through rocks opened a new world to petrologists. This is a porphyritic volcanic rock seen under polarized light ("crossed nicols"), including some large stripey twinned felspars and olivine (orange-pink) in a "sea" of much finer crystals. Modern equipment can investigate the chemistry of a single crystal.

ABOVE Living "tubeworms" covering Zooarium, a sulphide chimney named for its rich biota; Explorer Ridge, 2002.

RIGHT Deep-sea sulphide deposits: a manipulator arm holds a sample from a deep-sea chimney that has ceased activity, showing "fossil" vestimentiferan worm tubes. Such fossil deposits are now known to have a history of several hundred million years.

ABOVE A volcanic cone made of extraordinary carbonatite lava: Oldoinyo Lengai in Tanzania on the Great Rift Valley 120 km north-west of Arusha (*see p. 240*).
BELOW The remote Nuratau Mountains of central Asia, Uzbekistan, home to rare species of minerals—some of which have only recently been discovered—as well as to the Severtzov wild sheep (*Ovis ammon severtzovi*).

TOP LEFT Rare and beautiful mineral crystals from the Kola Peninsula, Arctic Russia: Kovdorskite, a magnesium hydroxy phosphate.
TOP RIGHT The cast taken from the original Koh-i-noor diamond, and (*below it*) the brilliantly recut version in the Crown Jewels.
ABOVE LEFT The rare and unusual 'wheels' of the mineral bournonite from Cornwall (chemically, it is copper lead antimony sulphide).
LEFT The Latrobe gold nugget (NHM 6020): 717 gm of crystallized gold from Mt. Ivor, Victoria, Australia.
RIGHT The "accursed amethyst" of Edward Heron-Allen, now hidden in a drawer in the Mineralogy Collections. Don't contemplate it for too long (*pp.* 246–47).

ARTISTIC TREASURES KEPT SAFELY IN LIBRARIES
ABOVE A confederation of owls from Audubon's *Birds of America* (1835–38), Plate 432; the species are *Athene cunicularia, A. noctua, Glaucidium gnoma* and *Asio flammeus.*
BELOW LEFT From Ferdinand Bauer's *Illustrationes Florae Novae Hollandiae* (1816): the gymea lily *Doryanthes excelsa.*
BELOW RIGHT Georg Dionysius Ehret (1707–70): *Senecio pseudoarnica* Less, or Seaside ragwort, sketch 11 from Ehret's original drawings of plants from Newfoundland. Joseph Banks started his botanical reputation with an expedition to Newfoundland in 1766.

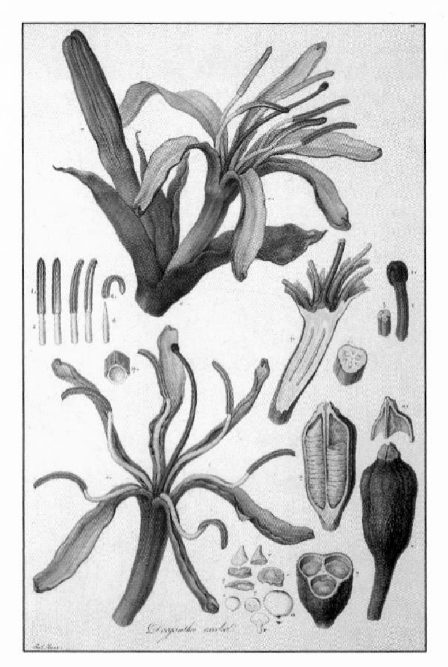

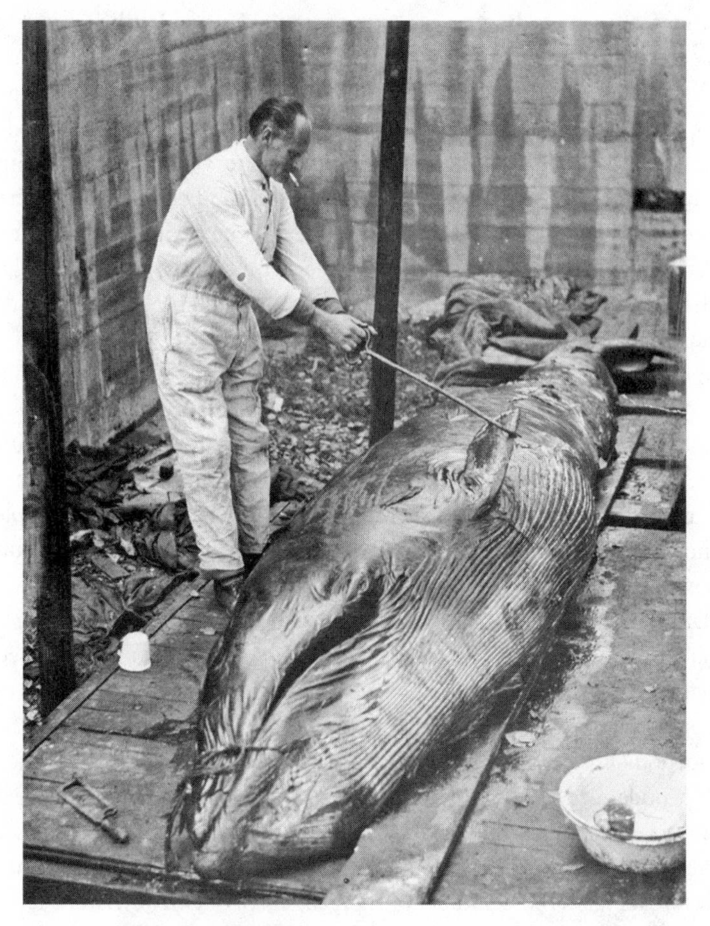

Purves working on a whale carcass in the late 1930s

little documentation, still drunk and late for a scientific engagement. He was arrested by the police, who fortunately found his letter of invitation from Pillari. In due course he was sent on his way. It is good to think that once upon a time this behaviour did not constitute the persistent and gross moral turpitude that would result in the removal of an employee from his post.

Whale research disappeared as a priority during the Museum's reorganization. Purves retired, a little early, but died at the age of eighty in 1995, thus providing an example to us all. The zoologists moved to the Darwin Centre, leaving behind only their molluscan colleagues where they had always been. I would like to take you into every office on the seven floors, for every expert in this building will have a story to tell, but that would take a book longer than this one. We will just poke our heads into one or two. Although an era of wood and brick has given way to one of glass, it is consoling to find that some of the curators' and researchers' offices still retain much the same feel as they always did. They are little nests lined with books and specimens, tucked away inside the glass exterior, homely as always. A microscope will stand on a bench, and maybe a stack of drawers from the collection. There is still that sense of continuity with the previous generations of taxonomists. I tend to see frequently the zoologists who work on the living crustaceans because these animals are included with my trilobites in the same Phylum Arthropoda. We are like second cousins. The crustacean people know more about living copepods than anyone alive. Professor Geoff Boxshall and Rony Huys are devoted students of these small, often planktonic crustaceans. This is no arcane study, for copepods make up a major component of the total marine biomass—all its life weighed together—and they are possibly even more important to the health of the sea than the much more familiar "shrimps" known as krill. The copepod *Oithona* may even be the most abundant animal on the planet. Copepods feed on tiny plankton, and in their turn provide the next important course in the food chain of the oceans. I recall seeing the sea almost black with them, swimming like so many animated peas, when I was working on the island of Spitsbergen, north of Norway. Arctic terns gorged on them, and the ground was stained pink with the bird droppings. Copepods are found in underground waters, even in temporary pools, and there is a mass of species in Lake Baikal—a "species flock" like the cichlid fish I described previously. New species are routinely discovered. They need to be made known to the world. There is evidence that copepods are sensitive to pollution and, being close to the bottom of the food chain, any decline they suffer will be mirrored in a fall in numbers of their predators. As with so many of the specializations I describe in this book, knowing about copepods is no obscure academic study: it's about feeding the world.

I will briefly greet the herpetologists, whose collections include pallid pythons packed like intestines into tall jars. Nick Arnold is the lizard man famed for work showing how lizard species coexist and recognize one another with curious dances, how different species are found on different islands, and how speciation events in turn relate to the islands' geological history. For the more general naturalist, Nick wrote the definitive herpetological guide to Europe, *Reptiles and Amphibia of Europe* (2003). Garth Underwood was the snake man and a kind of inverse Peter Purves, being a man of quiet sober dedication and reticence who continued to work with scarcely diminished vigour until his death at eighty-three in 2002. Or we might visit some people who study at the diminutive end of the zoological spectrum; let's call in to meet John Lambshead, who is a "nematode" man. Nematodes, or roundworms, are tiny but ubiquitous animals—any small handful of soil will have many nematodes in it. Unless you are a nematode man, I have to say that they have a somewhat limited aesthetic appeal. I usually see them wriggling across the field of view when I have been examining small fungi under the microscope. Furthermore, the parasitic habit has arisen on many more occasions in the nematodes than is the case in schistosomes. In nature, it is always an attractive option to bludge a living, as an Australian might say. Some are serious root pests. Another, but useful, nematode is the model organism *Coenorhabditis elegans,* which was so important in unravelling the genome and how it works. In the sea, nematodes can be found in great numbers in the mud anywhere from shallow-water habitats to the deepest part of the ocean trenches: so even a small grab sample can yield up plenty of evidence. The composition of the nematode population provides a very good indication of the level of pollution, for there are species that thrive under just about any nasty conditions, others that prove that everything is as it should be. John Lambshead has been much concerned with using nematodes to monitor the state of health of the sea floor, a matter that should be on the radar of any politician claiming green credentials. He is rather a dramatic man to be paired with such undramatic animals. With a ready smile, expansive gestures and hypnotic eyes, he is reminiscent of the Ancient Mariner, except that he "stoppeth one in three" to explain the importance of nematodes in understanding habitat diversity. There are

The author standing in a typical Museum office
(his own is rather less tidy)

actually far too many nematodes to name them all formally, and too few nematode men to do it, so a library of molecular sequences characterizing species as yet unnamed has been developed. It is quite practical to do environmental monitoring with a cast of characters for which the Linnaean binomial is still waiting in the wings.

The Darwin Centre encourages the public to come into the collections on special tours. Some of the scientists occasionally feel that they, too, are another set of specimens, laid out for inspection by amiably curious visitors. Regular lectures about the research work of the Museum in the Darwin Centre have been a success in opening our arcane world to scrutiny. The boffins come blinking into the light, and most of them love it. For their part, the audience like to see into the hidden world behind the galleries. Charles Darwin himself is not infrequently mentioned on these occasions—how can it be otherwise in this cathedral to his ideas, where many of his specimens still reside? No other scientist has given his name to so much. The capital of Northern Territories, Australia, is simply Darwin; Mount Darwin is the highest mountain in Tierra del Fuego. Several variations on Charles Darwin Research Institutes are scattered around the world. There are numerous *darwini* species names. He is celebrated in Darwin's finch, fish, frog, amphisbaenid, gecko, barnacle, sea slug, snail, beetle, cricket and the lowly thrip (and many more insects besides); two mice, one not seen in Galapagos since 1930, and the other a leaf-eared mouse under threat in South America; several spectacular fossils and numerous plants: all *darwini.* There is even a computer game called Darwinia. The endemic Australian shrub of the same name (*Darwinia*) is, however, named for Darwin's grandfather, Erasmus—not that you could tell from the name alone.

Newton or Galileo do not come anywhere near in profligacy of eponyms; perhaps the closest approach is Louis Pasteur in France, if you allow the inclusion of street names. What French town worth its name does not have a Rue Pasteur not far from its Rue Victor Hugo? And if Darwin has his -ism, Darwinism, then Pasteur has his -ization, pasteurization, although he has lost his capital letter in English. Is it conceivable that anyone would now say that they don't "believe" in pasteurization? It would be rather like saying they don't believe in germs or microbes, or even hygiene. Yet a comparable denial of Darwinism has apparently been growing in the last decades. Darwin and Pasteur were—very approximately—contemporaries, so those who deny Darwinian evolution are seeking to push back knowledge to some time before the 1850s. Imagine if it were medicine! I sometimes wonder whether those who disavow Darwin should also eschew the medical treatments that have been developed thanks to the insights evolutionary understanding has provided. I mean things like genetic disorders, or rapidly changing viruses like the one that causes AIDS, or cancer treatments arising from the discovery of oncogenes. Such unbelievers could still have most surgery with a clear conscience. As for the rest, well, I believe Thomas Culpeper had some good herbal remedies for the "bluddie flux."

5

Theatre of Plants

Laying out a plant for pressing is like laying out a corpse before a wake. Everything must be in its proper place: the arms by the side, the visage upwards showing the best face in the circumstances, the clothes perfectly arranged. The skilled plant presser moves with the trained efficiency of a mortician. Foliage is laid out just like a best frock to present its frills and furbelows; flowers are displayed to show their important features; roots are sometimes appended like a bundled tribute. The whole business is carried out automatically and without fuss. But, unlike the doomed cadaver, the laid-out plant is destined to endure, its spread-eagled best a voucher for as long as collections last. Everything is done to make sure that the specimen does not decay like its companions in field or forest. The paper on which it is laid out is acid-free, absorbent, of course, but superior to the kind of blotting paper we used as children. Wayward sprigs are held in place by paste. Covered by another sheet of paper like a winding-sheet, the plant will take its place in a pile destined for the flower presses. Some of these contraptions would be familiar to students of Linnaeus himself, comprising two opposing flat metallic plates between which the pile of sheets is sandwiched, the whole then being compressed by a mighty screw, the winding arm of which has massive brass balls at either end. The process was

Herbarium sheets laid out for inspection in the General Herbarium: preserved in perpetuity

first described in print by the Flemish savant Adrianus Spigelius in his book *Isagoges in rem herbarium* (1606). The end product is a herbarium sheet, which is often rather beautiful. I have to assume that this aesthetic sense is mostly an accident, but there is often something that recalls the old herbals about the layout of the plants on the sheet. The savants who prepared early floras of useful or medicinal plants used the woodcut plates illustrating their subjects as a demonstration of artistry. The plants seem to grow across the pages, but often in a somewhat formal way, as if to make the best use of the space available.

These practical works long predate Linnaeus, and their systematic arrangement is usually one of medical virtue rather than biological affinity. I have already mentioned Leonhard Fuchs, whose *De historia stirpia* (1542) is a landmark in the portrayal of flowering plants, for both accuracy and beauty, but some might regard the illustrations by Hans Weiditz in the work of Otto Brunfels as even finer. They include many life studies of plants found growing near Brunfels' home in Berne, Switzerland, where he was a Lutheran pastor. It would be difficult to

An illustration of *Clematis vitalba* from an old herbal of Leonhard Fuchs (1542) with the plant laid out decoratively upon the plate decide which of the genera subsequently named after these pioneers is the more showy: *Brunfelsia* or *Fuchsia.* The latter is certainly the more ubiquitous in gardens. John Gerarde's *Herball, or General Historie of Plants* (1597) is perhaps the best known of these early works, but the woodblocks made by Tabernaemontanus* yield more schematic prints that seem well on their way to becoming the kind of vigorous, twining growth featured on William Morris' wallpapers. I do not suppose for a moment that those who lay out herbarium sheets consciously follow the herbals, but there may be an instinctive tendency to use space in a satisfying way that does not "waste" it. There is one way to do the job that seems intuitively right. If we follow the democratic views of the German artist Joseph Beuys, it could be that a skilled preparator of plants just can't help making art.

Naturally, you cannot press a banana or a cactus or a tuber: these are popped in jars just like any toad or lumpfish.

Once hidden away safely in their folders, dried plants are great survivors. There are tiny malevolent creatures like booklice that will make a meal of glue, and damp is simply lethal to any stored dried material, but given a little care a herbarium is a safe place for a permanent repository. Some of the oldest specimens in the Natural History Museum are herbarium sheets that came from the original collections of Sir Hans Sloane, and so root back to the very foundations of the BM. The Sloane Herbarium includes a specimen of cacao (*Theobroma cacao*), a species that he described from his extensive natural history collections from Jamaica, where he had been the personal physician to the Duke of Albemarle early in his career. He appreciated the virtues of a drink made from mixing milk and chocolate: his own recipe, "Sir Hans Sloane's Milk Chocolate," was sold by Cadbury's until well into the nineteenth century. In this age of bijou *chocolatières,* it may be time to revive it. Sloane purchased or inherited a number of important herbarium collections at a time when the botanical world was just being explored from Europe. He acquired the first collections from the Cape in South

*As with Linnaeus, many of the authors or artists of the time took Latin soubriquets. Fuchs was known as Fuchsius, for example. Tabernaemontanus was Jacob Theodore von Bergzabern (in Alsatia), to which the curious name is a latinized reference.

Africa, a region now renowned as one of the most botanically diverse; he bought the then unique Japanese collection of Engelbert Kaempfer, who died in 1716; likewise the collections of James Petiver, who died in 1718, replete with North American species. He was voracious in acquisition, a true collector. He also outlived his rivals, which helps if you happen to have a bottomless purse. By the time he died as a very old man, his collections filled his manor house in Chelsea, which was then an almost rural part of London. In addition to the 100,000 specimens, 50,000 books and 3,500 manuscripts, and so on, that formed the nucleus of the British Museum, there were 334 herbarium volumes—Sloane's *Hortus siccus,* literally his "dry garden." These have been moved, and moved again through the several transmogrifications of the BM. The first move was from Sloane Manor to Montagu House in Bloomsbury in 1756—which became the original British Museum. In the nineteenth century, as the collections continued to grow, the herbarium was moved to a room in Smirke's new grandiose neoclassical edifice, where the antiquities and drawings remain today, and where the Elgin Marbles still draw the crowds. Finally, the famous volumes were transferred to Waterhouse's purpose-built extravaganza in South Kensington, and long may they remain there.

The Sloane Herbarium has survived better than most of his other collections (but see p. 21); specimens mounted on paper truly do defy time and decay. Two hundred and sixty-five volumes are now housed in a Special Collections room, with controlled humidity and temperature and filtered air—they have never been so well cared for in their quarter-millennium history. As for his zoological specimens, they have fallen foul of carpet beetle and moth. The stuffed birds and mammals have been eaten away—even the eighteenth- and nineteenth-century "preservatives," which were mostly poisonous compounds of arsenic and mercury, probably did more harm than good. It was reported that in 1806 "most objects of the Sloane collections were in an advanced state of decomposition, and they were buried or committed to the flames one after another. Dr. Shaw [then the man in charge] has a burning every year; he called them his cremations." Thus passed from the world one of the founding collections of the British Museum, gobbled first by larvae and then by fire. The curators had no choice in their actions, as they did not wish to infect the rest of the collections. A nervous awareness of pests has been a preoccupation of the Museum staff ever since.

The word "botany" may call to mind flower presses, and earnest ramblers bending over a small weed in a nature reserve, but it has a much wider compass than just flowering plants. The Botany Department has a whole floor devoted to cryptogams. This is one part of the Natural History Museum that has hardly altered since I joined the organization. Tucked away in the attic at the east end of the Museum, it comprises an open gallery under a low roof with many bays surrounded by old polished cabinets. Curators are sequestered away in the bays like grubs inside a peach: they can be found if you are sufficiently determined to winkle behind the bookcases and piles of folders, but you feel that they wriggle into the light reluctantly. "Cryptogam"(Greek: "hidden marriage") is a broad term for plants that have "hidden" reproduction, rather than replicating themselves from obvious things like seeds. It is an old term, going back to the time of Linnaeus, when it was still unrecognized that plants could reproduce from minute propagules like spores—which often look to the naked eye like nothing more than brown dust. It seemed against common sense that something so minute might have the potential for producing progeny. The development of the modern microscope allowed a different view. So here in the Cryptogamic Herbarium there are the experts on ferns, mosses, liverworts, algae and lichens: these are the organisms that decorate branches and stream banks, paint rocks in every colour in the palette sold by Windsor and Newton, and grace rock pools with fronds and feathers of red, brown and green weeds that toss with the tides. To the rest of the Museum such plants are known simply as "Crypts." The Crypts have their own library, and their own ways.

For many years this part of the Museum also included experts on fungi, the mycologists; however, the appropriate collections were transferred to the Royal Botanic Gardens at Kew in 1969 and 1976. I have always felt a particular connection to John Ramsbottom, who was Keeper of Botany for an astonishingly long period between 1930 and 1950, and was also a distinguished mycologist. Ramsbottom wrote one of the celebrated New Naturalist series: *Mushrooms and Toadstools* (1953). As a young naturalist I was entranced by his blend of esoteric learning and scientific expertise; a mixture of scholarship, history and science which seemed to make nonsense of C. P. Snow's notion of the "two cultures"—arts and sciences. Ramsbottom recognized only one seamless culture, and could combine quotations from Pliny or sixteenth-century herbals, or lines from little-known books of poetry, with the discoveries of twentieth-century microscopists or chemists. From his writing I intuited a great depth of scholarly knowledge combined with an appreciation of fungi in their natural habitat, and a rather dry humour buried somewhere below that. There is a diverting preface from the editor of the New Naturalist series noting that the original manuscript submitted by John Ramsbottom had been twice as long as the version finally published and "even then, he complained that much had to be omitted." The final book is replete with footnotes, and sections set in very small type. I can easily imagine the battle between editor and scholarly author, the one for trimming, the other for expansion, and the resulting selection of compromises. I am glad that the cutting tendency did not eliminate some curious little footnotes, such as one that records, "At Clerkenwell Police Court in March, 1937, the defence argued that a man accused of being drunk in charge of a motor car was suffering from the effects of a rather liberal meal of mushrooms." He does not record how this went down in court, although it sounds like a useful defence to have up one's sleeve should the need arise.*

Oddly enough, although mushrooms decay fast, they make rather good permanent collections. When they are dried rapidly over gentle heat, they lose their water but retain their microscopic features, such as their spores and the fine structure of the gills, which are particularly important in identifying species. They can be tucked safely into voucher envelopes and curated in collections. Colour tends to fade, and it has been a tradition of mycologists to make watercolour drawings of fresh material before the drying process takes effect. Nowadays, colour

*The quality of colour reproduction has improved markedly since Ramsbottom's time, and the photographs in the original edition look rather unnatural to modern eyes. So after half a century it was time to have a replacement, and this has been published by two of the current Kew mycologists, Brian Spooner and Peter Roberts. The science is more up to date, of course, but one misses the footnotes.

photographs provide quicker and easier records, although they can sometimes fail to capture the "soul" of the fungus that a skilled watercolourist can pinpoint. Lichens are easier to dry than fungi, because most of them spend a large part of their existence in a desiccated condition. They form crusts on or even within rocks, or crispy daubs on tree branches, or dangle from twigs like wispy grey-green beards. They are the toughest organisms on Earth. I have seen a dozen species painting rocks green, white and orange in the highest regions of the Arctic. They relish the most exposed faces on granite tors on Dartmoor. Lichens are a symbiosis between a fungus and an alga,* joint partners in the business of survival. When little coloured cups appear on the surface of the lichen, this is the fungus showing its face during reproduction.

Although the fungi collections moved to Kew Gardens, the lichens stayed behind in South Kensington, along with the other Crypts. They reside in their folders in the polished cabinets up in the "gods" of the Natural History Museum. Their systematic study goes back to Erik Acharius, a marvellous artist, who divided the original single lichen "genus" into forty, and thereby alerted the botanical world to their intriguing variety of form. For many years of my career the lichen man in the Natural History Museum was Peter James, diffident master in the art of their identification. With his trim beard, half-moon specs and amiable if distrait manner, he was almost exactly what a member of the public might imagine a museum academic to be like. There are more lichen species than you can shake a stick at, most of them passed by unnoticed in the field even by the most broad-minded naturalist. The British Checklist lists 2,272 species of lichens for these islands alone. Peter's name is still spoken in hushed tones by the lichen community, since he has a preternatural ability to identify the most obscure species from memory. His successor, William Purvis, is as devoted to the cause of these under-studied organisms.

Lichens are particularly important as indicators of pollution because they readily absorb heavy metals into their tissues. They simply mop up elements like lead and cadmium. Lead was formerly present in

*Strictly speaking, the "vegetable" partner is termed a dinoflagellate in many examples.

appreciable quantities in petroleum spirit. In Britain it was practical to assay the damage done to the environment by mapping lichen species; most are unable to tolerate lead pollution for long, but those that can proliferate at the expense of others. The species plot out the state of the environment. It is extraordinary to see how precisely the lead pollution traced the course of trunk roads, forming a series of "corridors" with low lichen diversity criss-crossing the landscape. However, at least the lead content falls away laterally from the roads themselves. It is a wise precaution not to eat mushrooms picked by main roads, as high lead content has been associated with loss of brain function. The introduction of lead-free fuel received a boost from such findings. Far wider regional pollution across industrial areas is also faithfully recorded by these humble living patches. Inner cities can be very low in species as a result of the pollution from the "dark, satanic mills" of the last 150 years. As you move westwards across the British Isles, the number of lichens increases. The dominant weather systems move in from the clean Atlantic Ocean, flushing most pollutants eastwards, and the lichens revel in the moister atmosphere in the west. In the old oak forests in North Wales every boulder is dappled with lichens, while the twisted oak branches are heavily draped with leafy and feathery forms. Some of these lichens yield vegetable dyes with lovely natural colours, yellows especially, so in addition to dressing the trees, lichens might well help to dress us, too. Around the old mines on the hillsides nearby, other lichen species take up noxious elements or pollutants, for which they seem to have a particular affinity. These can now be accurately assessed using modern mass spectrometry (see pp. 236–37). They are useful things, these unspectacular lichens, for diagnosing the health of the planet. As we have seen so often before, an apparently arcane expertise in an organism that would be passed by unnoticed by many relates to important contemporary issues.

Many lichens also grow very slowly, some just a few millimetres a year, and some even less than a millimetre. It is an interesting problem how to determine this rate of growth. One way to estimate the average growth rates is to use gravestones. These memento mori are a favourite habitat for lichens—indeed, there are some species that are now hardly found outside old churchyards. This is partly because many of these sacred acres have escaped chemical spraying or artificial fertilizer for decades or longer. As we have seen, lichens are unusually sensitive to all pollutants, even those of alleged benefit to farmers. Some churchyards are little patches of medieval habitat. Lie back beneath an old yew tree, and imagine priests and squires and villeins going about their Sunday business. A gravestone includes that very useful piece of information—a date. When erected they are pristine, but soon time and lichens make their mark. Lichens on flat gravestones tend to grow outwards in a regular circle, so the diameter of the circle is proportionate to its age. The largest circle found on a gravestone of a particular date will give an approximation to the maximum rate of growth. There will be a certain range of variation as a consequence of local conditions, and variation in the time of first colonization. Furthermore, as time passes, new species of lichen will join the gravestone habitat—and younger rings will "cut" through older ones as they grow, so revealing the order of succession of colonization. A good gravestone will accordingly yield a complex narrative, and many gravestones will provide usable statistics. Rates can then be applied to other sites. The use of lichens in dating is known as lichenometrics. Although not without its critics, lichenometrics has revealed some interesting figures. It seems that lichen growth rate has speeded up in high latitudes since the industrial revolution, and that this may be connected with global warming. Because they can be found nearly everywhere and grow so slowly, lichens are potentially a ubiquitous biological "diary" that records changes to the environment and atmosphere on the century scale. They are the Crypts' answer to the tree rings on that ancient giant *Sequoia* trunk on display outside the Botany Department.

As with just about every other field of study mentioned in this book, the molecular revolution has forced us to rethink ideas about the relationships of lichens. Molecular evidence shows that fungi have become "lichenized" on many occasions: to put it another way, they have repeatedly made a contract with algae to enter into their special partnership. The great majority of such fungi are "waxy cup" ascomycetes (see pp. 66–67); this is easy to see on those lichen species that bear little red or yellow cups on their surfaces, because they are not very different in appearance from the spore-bearing fruit bodies of their relatives, which can be found on forest paths or compost heaps. Next time you are in a country graveyard, peer closely at a headstone and you will be almost sure to see these little coloured cups or plates on an encrusting lichen. A few gilled, mushroom-like fungi have also taken this curious evolutionary path. On the other hand, the "algal" partners of the lichen symbiosis are identical to common species living free in nature. It seems that it must have been the fungal partner that hijacked the algae into the collaboration to become lichen. The fungi obtain their carbohydrates from the algal, or blue-green bacterial, partner. The latter employ photosynthesis to manufacture carbohydrates in their cells, at the same time "fixing" atmospheric carbon dioxide. For their part, the fungi access mineral salts like phosphates that would not otherwise be available to the algae. They live on rainwater and dust. Working together, these two organisms produce a collaboration that is remarkably tough. No doubt lichens will outlast us all.

I have dwelt on lichens in this chapter because I believe that symbiosis is a telling metaphor for the way that scientists work in a national museum. There is a close relationship between scientist, curator and librarian. Like the "algal" partner in lichen, all three are capable of independent existence. But working together in the Museum environment, the different virtues of all three partners produce something greater than any individual could on his or her own. The scientist might have the kind of manic devotion to research I have already described several times, but without a great library the work might not have the depth that makes a classic; the curator for his part is vital in ensuring that the contribution is there for posterity, and for other students. The collections and the work upon them should last like a crusty old lichen on a storm-blasted rock. Recall Sir Hans Sloane's voyage to Jamaica in 1687. The specimens are still safely curated.

Molecules also confirm that fungi are not plants. The old classification on which the museum collections were organized lumped all the Crypts together. Now that the molecular trees have helped us peer into the deepest parts of evolutionary history, we can understand that the fungi split off from all the rest of living things very early in the history of the planet, perhaps as long as two billion years ago—the exact date is still the subject of dispute. As a whole, fungi are closer to the protozoan line that leads to animals than they are to the branch that leads to algae and, ultimately, to ferns and flowers. Lichen is therefore a collaboration between two kinds of organisms destined to have completely separate trajectories through Earth's long history.

It is a little ironic that the fungi left their animal (second) cousins at the Natural History Museum and went to join the higher plants at Kew Gardens, while their symbiotic partners stayed behind. This divorce of natural partners was a consequence of the Morton Agreement of 1961, so called because Sir Wilfred Morton was then the chair of Trustees. The Agreement sought to divide out the taxonomic responsibilities of the sister institutions—no point in duplicating research, or so the argument ran. The world was actually carved up into fiefdoms. For example, according to the official document, the BM would have "Northwest Africa from the Atlantic Islands to Tunisia; S. Tomé and other W. African islands, excluding Fernando Po," whereas Kew might have the "rest of Africa including Madagascar and Mascarene Islands." South America belonged to Kew, while the arctic areas and Britain belonged to South Kensington. It did seem rather proscriptive, to say the least. But it worked in a kind of ramshackle, compromised way, even if it did separate the two halves of the lichen. The British responsibility at the BM also resulted in some major achievements, such as *The Island of Mull: A Survey of the Environment and Vegetation* (1978), which probably remains the most thorough survey of any biologically rich area of the British Isles, at 5,280 species, no less—including the fungi, despite the limitations of the Morton Agreement. Mull now has several nature reserves, and, being on the wet west, is full of lichens.

The Crypts also helped to save the United Kingdom during the Second World War. At the outbreak of war there was a curator of "seaweeds"—marine algae—called Geoffrey Tandy. He had come to the Natural History Museum from the BBC. He had a wonderful voice, and was a close confidant of T. S. Eliot; he it was who made the first ever broadcast of *Old Possum's Book of Practical Cats.* According to his successor, Tandy was a competent taxonomist and contributed many herbarium sheets to the collections, which are still referred to today. However, he was not much of a publisher, and wrote only two scientific papers while at the Museum, a deficiency that eventually led to him being called before the Trustees; he excused himself on the ground that "writing up" was not part of his job description. He apparently ran two families in tandem, from which one son on the illegitimate side survives. The reason he saved the country—possibly even the world—from Nazism was because he was a cryptogamist. Evidently, a functionary in the Ministry of War had never heard of cryptogams, and thought that Tandy must have been an expert in cryptograms (that one extra letter ensures that the words appear next to one another in the dictionary). He was recruited to Bletchley Park—centre of signals intelligence during the war—because of his alleged talent in solving messages written in code. He had to work alongside the great brains that were tackling the mysteries of the Enigma Code—the only seaweed man among the ranks of cryptographers. It was a most fortunate screw-up. When sodden notebooks written in code were recovered from German U-boats, they seemed beyond recovery. However, Geoffrey Tandy knew exactly what to do. The problem was actually not so different from preserving marine algae. The Museum supplied the appropriate absorbent paper, and the pages covered in cryptic language were saved from soggy obscurity. The code was cracked, thanks to the fact that the word Linnaeus used for organisms reproducing by spores was but one letter different from the word describing messages written in code. One thinks of James Thomson ("The Seasons," 1730): "A lucky chance that oft decides the fate / of mighty monarchs." Or dictators.

We seem to have arrived once more back with Linnaeus. Since binomials given by Linnaeus represent the standard for plant names, it is obviously important to know exactly what they mean, both to understand how Linnaeus looked at the natural world and to fix the species in nature to which his names refer. It may seem strange that this still needs to be done, three centuries after Linnaeus' birth. However, botany, like other disciplines, grew up haphazardly. There was plenty of opportunity for muddle and confusion. Although material was exchanged between scientists, this did not happen as widely as it does today. Because many parts of the world were being explored for the first time, the circulation of illustrations was often all that was done to make new discoveries known to waiting European savants. Foreign travel was difficult and expensive, so that scientists in the eighteenth and nineteenth centuries could not just pop over from Germany to consult the Sloane collection without the help of some serious patronage. A few of them were privately rich. Illustrations were often superb, but equally they could be inadequate and sketchy. Nor did Linnaeus necessarily have a specimen to hand when he coined a name. He often referred to illustrations in older floral works and simply dubbed them with his binomial. This chapter takes its title from one early work, *Theatrum Botanicum, or, Theater of Plantes,* by John Parkinson (London, 1640), which seems an appropriate label because so many of the stories from the plantsmen are thoroughly theatrical. I have already mentioned publications by Brunfels, Fuchs and Sloane, but there were many more. I particularly like J. J. Dillenius' 1732 *Hortus Elthamiensis,* an account of the garden of Dr. Sherard in Eltham, Kent,* because it combines a feeling of exoticism with the name of an archetypal London suburb. Linnaeus himself published *Hortus cliffortianus* (Amsterdam, 1737) with illustrations prepared by the incomparable G. D. Ehret. These publications, for all their splendour, exude a whiff of vanity publishing among the rich—"my garden is more exotic than your garden"—but they endure as works of art as much as science.

Another source of confusion is that original collections often included more than one species filed away under a single name. In this case a selection of *one* type specimen has to be made from the several specimens available, and preferably a wise choice that stabilizes the name as everyone had used it for centuries. Then, in the early days, as always in science, there were personal animosities that encouraged one worker simply to ignore another. The same plant may well have been named more than once as a result. Human folly has always been built into the system, and the result can be a mess—or, to put it more precisely in this case, taxonomic confusion. What was evidently needed to stabilize Linnaeus' names for plants was a series of type specimens deposited in recognized and properly curated herbaria on which his names would be pinned for ever. Thus was initiated the Linnaean Plant

*William Sherard founded (I should properly say endowed) the chair of Botany at the University of Oxford in 1734, and Dillenius was the first Professor. The holder is still known as the Sherardian Professor. Readers in the United States might also like to know that Bob Hope was born in Eltham.

Name Typification Project in 1981. About time, too, one might think, finally to fix the names of our dahlias and daisies. The whole project is co-ordinated from the Natural History Museum, and the man in charge is Charlie Jarvis, an unfailingly amiable inhabitant of one of the cubicles off the sides of the main herbarium. He is as surrounded by piles of ancient leather-bound tomes as anyone in the Museum. In fact, his small piles of large old tomes have bigger piles of small old tomes piled on top of them. I cannot imagine Charlie Jarvis giving me the poisonous snarl I received from one of his predecessors when I made my first peregrinations around the general herbarium. He has to be a combination of lawyer and diplomat, because Linnaean names are attached to flowering plants from all around the world, and not everyone always agrees on the most common-sense conclusion. There are nit-pickers and hidebound legalists to contend with, and also those who measure their stature by their recalcitrance. Charlie has to suggest what specimen in which particular collection both conserves Linnaeus' concept of a species and stabilizes the meaning for everyone else. I suppose that what I am implying is that Charlie Jarvis is some kind of saint.

The publication of *Order Out of Chaos: Linnaean Plant Names and Their Types* in 2007 marks the triumphant completion of the project: all 9,131 of Linnaeus' names that map out the beginnings of the chart of the botanical world are now at last properly fixed for future generations. Every name has its identity hitched to a type specimen. This is the opposite extreme of the kind of science that chases a result in the short term to make a big splash in the *New York Times.* But is there not something enormously admirable about a project completed after twenty-six years of scholarly research? To give an example from one small part of the work: Linnaeus described and named twenty-seven species of the genus *Solanum* from the New World. *Solanum* is the genus that includes the tomato and potato, so it is of considerable importance commercially, especially now as genes of its close relatives are being investigated in case they confer natural immunity to disease. Obviously it is necessary to know how many species there are, and what to call them, before anyone can get on with the business of plant breeding. In 1990, specimens were selected as types for all the Linnaean species. It is hardly surprising to learn that many of these types were chosen from the

Linnaean collection—the herbarium sheets that once belonged to the master. What may be more surprising is that these sheets are not in the University of Uppsala, Linnaeus' home campus in Sweden, but in the middle of the west end of London, in Piccadilly, just opposite Fortnum and Mason's—the famous grocer. As visitors pass through the entrance facing the Royal Academy in Burlington House on the north side of Piccadilly, on the other side of the road from Fortnum's, they might notice on the left a much more modest entrance to the Linnean Society of London.* In the basement of the Linnean Society there is a well-protected and air-conditioned store that contains the herbarium of Linnaeus himself. It has a door like that of a bank vault, appropriate enough for the treasure it contains. The collection fetched up in London because Sir James Edward Smith bought it from Linnaeus' widow for the sum of £1,050 in 1783. He certainly got a bargain, regardless of those tedious calculations that tell you how much that money would be worth today, for he had purchased something timeless. Because of inefficiency or indecision, the Swedish government did not make up its mind quickly enough to purchase the collection of its most famous scientific son, and Smith stepped in. The Swedes realized their mistake and dispatched a man o' war to try and overtake the collection as it sailed on its way to England. Fortunately for the Linnean Society, the collection got away. I doubt whether an herbarium has ever before or since been the object of a diplomatic incident. The collection formed the nucleus around which the Society grew from its foundation in 1788, with Smith interminably as its President. From this famous collection the types of the Linnaean tomatoes were selected.

The Linnean Society is the "club" to which many Natural History Museum employees belong. It is the normal place for biologists who study whole organisms to meet; it is also the oldest biological society in the world. It was established for "the cultivation of the science of natural history in all its branches," as its Charter states. The Geological Society lies opposite the Linnean on the other wing of Burlington House,

*The Linnean Society spells its name without the extra "a" of Linnaean, which confuses me, as no doubt it does others. I was told that this is because the Society is named after Linné (his Swedish name) rather than the latinized version in common use.

One of the sheets from the herbarium of Linnaeus in the Linnean Society of London. This is larkspur, *Delphinium consolidum.*

and has a similarly distinguished history in the earth sciences. Just around the corner are the astronomers and the antiquaries, and the chemists are also nearby. An agglomeration of learned societies has no collective noun, although a "disputation of societies" comes to mind. However, I do not suppose that any current Fellow of the Linnean Society would dispute that the paper presented on 1 July 1858 represented the most important occasion in the Society's history, for this was when Charles Darwin and Alfred Russel Wallace jointly presented the "first draft" of the Theory of Evolution. These few minutes reset the agenda for biology, and indeed much of society; and, of course, changed the meaning of all those herbarium sheets now stored so carefully beneath the street in bustling Piccadilly. The many *Solanum* species, sharing their distinctive pointy flowers and succulent fruits, were not specially created each to its own design but were descended from a common ancestor. Fundamental resemblance implied evolutionary relationship. So it was possible to work out how organisms were related one to another: this goal of describing *phylogeny* followed logically, and thereby set a research programme for research scientists at the Natural History Museum and its sister institutions around the world.

The President of the Linnean Society would preside over a different body after 1858, even if the faces before him were the same as before. I should add that he would preside from what is almost certainly one of the most uncomfortable chairs in the world. It is made of crocodile hide, and I have watched successive Presidents wriggle uncomfortably about on it in the famous if incommodious lecture room, while portraits of Darwin and Wallace look sternly on. By contrast, the library upstairs is comfortable, bright and spacious, its two floors lined with books and attractively classically columned, giving on to an airy atrium with bays commanding a view of Piccadilly. There is much to entertain the natural history bibliophile on the crowded shelves. The library is just the place for colleagues to gather to discuss the evolution of daisies or the parasites of fishes over a glass of wine after the formal presentations downstairs. The President will mull over business with the Secretaries. Hobby-horses will be ridden, bees will buzz in bonnets. Brian Gardiner, formerly Professor of Zoology in London, might well be pressing the case for the relative neglect of A. R. Wallace compared with the near sanctification (in an agnostic kind of way) of C. Darwin. He has a point. A special fund was set up to rescue Wallace's monument from overgrowth and collapse, whereas Darwin's house and environs at Down are applying to become a World Heritage Site. Every aspect of the great man's life has been the subject of biography, and usually several. Quite often somebody will be patiently explaining to someone else why their own particular organism is the best way to understand this evolutionary process or that.

The expert in the Natural History Museum on *Solanum* and its relatives is Dr. Sandra Knapp. She occupies another of the alcoves in the General Herbarium, one on the other side of the vault from Charlie Jarvis. It is difficult to describe Sandy as occupying any particular space, since she moves so frenetically around the Museum and seems to recruit all available spaces unto herself. She is one of a growing number of Americans on the staff in London—which is a measure of the increasingly international stance of the scientific research. She exemplifies the very best qualities of her nation—dynamic and enthusiastic, absolutely devoted to the cause of floras, good humoured, and completely without regard for the more ludicrous side of institutional life. Like me, she spends much time rooting around in piles of papers looking for something that has gone astray; unlike me, she usually finds it. Although now touched with grey, she is effectively ageless, so long as she can tap into the secret supply of energy to which she has exceptional access. The flowering plant family that includes the tomato is known as the Solanacea, and comprises about six thousand species belonging to about ninety genera: that is a lot to know about, *pace* the beetles. It is sometimes known as the nightshade family, because it includes the poisonous deadly nightshade, *Atropa belladonna,* which was also one of the first effective medicinal plants described by the old herbalists, although its specific name refers to its other use as a pupil dilator, allegedly to make women more attractive: *la donna è bella.* The species included in Solanaceae range in form from small trees with big leaves living in rainforests to dry, spiky herbs that are confined to deserts. The family is almost globally distributed.

Sandy's work has been directed particularly at the genus *Solanum* itself, which is one of the most diverse genera of the plant world, for it embraces something like 1,500 to 2,000 species, including some of those most important for the nourishment of the human race. I exclude another solanacean plant, tobacco (*Nicotiana*), from this role. Sandy has done a great deal of fieldwork in the Andes in search of new or forgotten *Solanums,* and nothing gives her greater pleasure than making an unexpected discovery on some remote mountain slope. She boils over with enthusiasm at the mere prospect of a forthcoming expedition. Many of these out-of-the-way places are hardly known botanically. Members of the tomato family are not inconspicuous little weeds for the most part, but Sandy has discovered and named fifteen species new to science, as well as clarifying the meaning of many previously named varieties. In the process of collecting her particular plants, she has also collected many more belonging to other families, and this has resulted in the naming and description of *dozens* of new species. Since the possible medicinal properties of these plants are unknown, this gives a very practical meaning to botanical exploration. Who knows if some small herb may not cure the world of its outstanding ills? Not that I subscribe to the utilitarian view that plants are only good for what we can get out of them—it should be enough to add another beautiful (or even plain) item to nature's inventory. We need to know what there is in the world for us to look after, regardless of its potential use. It almost goes without saying that molecular sequences have been extracted from these *Solanum* plants to reconstruct their evolutionary history. The tomato (*Solanum lycopersicum*) is one of the few plants to have its genome well on the way to being sequenced. There are projects afoot to work out how the tomato genome can be improved and made more disease-resistant in an environmentally sensible way. These include SOL, the International Solanaceae genomics project, which is promoting and collating all the new molecular information from the family. But everything in research depends on having a sound taxonomy of the plants concerned. If you don't know what to call it, you don't know how to study it.

For many years Sandy has been organizing and contributing to a huge co-operative effort to describe and compile the total flora of the central part of the Americas—including the southern part of Mexico and the central American republics. This is the first major flora in Spanish designed to help local botanists with recognizing and collecting their flowers and ferns, and will be a benchmark in the conservation of little-known species. It is called—accurately, if inelegantly—*Flora Mesoamericana.* As a model of international co-operation it is exemplary; besides the Natural History Museum in London, the co-organizers are the Universidad Nacional Autónoma de México, and that most distinguished of United States conservation bodies, the Missouri Botanic Garden. The seventh volume has just been published, but it will eventually run to ten volumes. The scholarly works are published alongside field guides for the use of native peoples and local botanists. This should go some way to breaking down suspicions of botanists as "thieves" coming in to "steal" the herbal secrets of endemic peoples. *Flora Mesoamericana* follows a long tradition of major floras. Long before I joined the Natural History Museum, a great work on African plants, *Flora Zambesiaca,* had begun in 1960 and now extends to thirty-two published volumes. According to the arcane formula of the Morton Agreement this work is Kew Garden's responsibility, and remains so, but this did not stop Edmund Launert spending much of his working life upon it as a Museum staff member. Edmund is a small rotund German with a penchant for cigars, which kept the Senior Common Room bathed in fragrance in my early days. He has managed to write standard works on orchid cultivation, medicinal plants and scent bottles—on which he is an authority—as well as contributing to major floras and being the world's expert on African grasses. He is also a marvellously humorous storyteller, to which a relict German accent adds the final touch; more than one story in this book owes its origin to Edmund.

Floras go back to the early days of botany, although as we have seen they were originally accounts of the plants of famous gardens as much as of regions. *Flora Mesoamericana* breaks new ground in exploiting the internet to make it available to those who wish to use it in the field. There is a special online version provided with multiple indices so that the confused botanist can soon find a source of illustration for the species before him, and the correct name for it. On a more modest scale, field guides to the plants of particular botanical "hot spots" are useful to many amateur devotees of rare and interesting flowers. Islands in particular have endemic species, many delightful, and some in need of protection. Sandy's colleague Bob Press has compiled a flora of Madeira, for example, which helps the field botanist with recognition of that island's many special plants. In my view these are as important a contribution as specialized floras, all the more so because they broadcast our expertise so widely. My field guides to the British flora are the most dog-eared of all my books, and if wear is proportionate to affection, they must also be the most loved.

With the spread of internet storage and access for natural history information, there is an important issue about what constitutes publication, and even whether we need conventional books at all. There is no question that the web is an extraordinary resource for botanists, zoologists, entomologists and palaeontologists. If information can be freely posted on the web, the question needs to be asked whether it is necessary to buy a book at great expense that merely repeats the same information in printed form. Do trees have to be sacrificed for nothing? Identification does not require fat manuals. Botanists in particular love keys: dichotomizing keys. These have a very long tradition in botany and mycology, and I rather like them myself. For example, if you know the genus of the plant before you and want to determine the species, you need a key to that genus; it might begin: flowers blue *or* flowers red—your first choice. Once you have decided that the flower in front of you is red, you then go to the next series of choices, which might be: leaves entire *or* leaves divided. Leaves entire—good! Next choice: stem smooth *or* stem hairy—undoubtedly the latter. By now, you might have arrived at a species name—the one with red flowers, entire leaves and hairy stems—say *Rubrifloria hairystemmia* (to make one up). We could have finished up with a species with blue flowers, divided leaves and smooth stems—say *Coeruliflora glabristemma* (to make up another one). The dichotomizing key really is quite simple—in theory. However, if the species in question belongs to a genus like *Solanum* with more than a thousand species, negotiating one's way through a key can become quite tricky. It is very easy to take a wrong turn somewhere and finish up with the wrong name. It's rather like missing a vital turning off a trunk road and finishing up at Peebles, or in a DIY superstore. But with a little practice a key is a great asset—and it is ideally suited to placing on the web.

There are already dozens, if not hundreds, of such keys available there. I find this development entirely positive. It takes the scientist out of his turreted redoubt in one of the remote corners of Gormenghast and places him or her at the disposal of all interested parties around the world. It passes on expertise acquired through hours of burrowing through obscure tomes so that anyone can use it. This is a kind of democracy of learning, a generous gift from the cognoscenti to those who wish to learn. But it does not invalidate the field guide or the monograph. It will still be a pleasure to take a pocket book into the flower-laden roadsides and hills in Madeira for consultation. You *could* take a computer, and summon up the data stored in the great virtual mind, but—if one is honest—it is easier to take out a pocket book from the anorak and leaf through a friendly field guide. Or, for the definitive illustration of a plant, surely the old skills of the botanical artist still have a place. I would hate to think that the sublime artistry of a latter-day Ehret could be outflanked by an efficient digital camera. I hope that this is not mere sentimentality or artistic snobbery on my part. Nobody could dispute that colour photography of wildlife has improved enormously—good photography has become an everyday achievement. But this quotidian competence cannot be a substitute for the exquisite productions of a botanical artist, where the essence of the plant is distilled into a plate with all the assurance of a Van Dyck nailing character into the span of a canvas. Let us celebrate both the advances in technology that make identification of organisms a common right, and the esoteric skills and experience of specialists that deserve to live on.

Where the web will have the greatest impact is in making available technical information and illustrations to experts and amateurs around the world interested in particular animals or plants. This applies across the Natural History Museum, and not just to botany. You could imagine a field botanist in base camp in the jungles of Indonesia having a hunch about the identity of a particular plant—and instantly conjuring up the type material on his computer to make a comparison there and then. This is one way to turn a national museum into a truly global resource, and, thanks to their long history of exotic collections, one with particular use in the developing world. My palaeontological colleague Andrew Smith is compiling a catalogue available on the web of all the genera of sea urchins (echinoids), including, naturally, many known only from fossils. There are beautiful photographs of museum specimens from several views, descriptive notes and accounts of geological occurrence. The numerous "hits" on the site from all around the world testify to the usefulness of such an online resource. The arachnologists—spider people—have a web-based guide to European species, while Norman Platnick from the American Museum of Natural History is compiling a global catalogue of more recent spider literature. The list of useful guides, catalogues and keys on the web is getting longer every day. Type almost any scientific name into Google and an image will be recruited from somewhere or other. What these images do *not* necessarily have is the imprimatur of somebody who really knows their stuff, because there is little quality control on the identifications placed on the web. Some of the identifications could well be wrong, and thus do more harm than good. The role of the expert will more and more be to validate such websites. Of course, even experts get it wrong from time to time, but an advantage of the web is that corrections and corrigenda do not have to wait for a paper publication.

However, most scientists would agree that for a "science of record" like systematics it is still desirable to have paper copies stored safely away on library shelves. There are good practical reasons for this, which go beyond just an undeniable feeling of permanence. The trouble with computers is that they deal best with one or two images at a time. When the systematist is plying his or her trade the office tends to get into a state that might be kindly described as creative chaos. This used to irritate a former Keeper of mine, famous for his orderliness, who would declaim "untidy desk, untidy mind" whenever he visited my office. I never had the courage to reply "empty desk, empty . . ." What he was referring to was piles of books, many of them open at illustrations of trilobite specimens. When we make visual comparisons, we do not do it one image at a time; the human eye and mind are adept at making multiple comparisons, at weighing up similarities and differences. So if there is an unidentified fossil or plant to hand, scanning by eye around maybe ten publications (one or two strewn on the floor) is quite the best way to make comparisons. A good systematist is often described as having a good "eye" for a character—which means recognizing the significant features and filtering out the rest. The heap of books is not likely to go away soon, unless economic factors force "real" publication into extinction.

There is something about being a botanist that encourages sex. It would be fanciful to suggest that this harks back to Linnaeus (yet again) for his success in classifying plants according to the sexual system—his naming and numbering of the parts of the flower, stamens, stigmas, styles, ovaries and so on. Early objectors to the Linnaean system thought that looking at plants in this fashion might encourage lasciviousness, or else that the system was not proper for the eyes of ladies to contemplate. This does seem strangely consistent with some of the more arcane parts of the Kinsey Report in which a small percentage of the population were reportedly aroused by chrysanthemums. However, the exclusive use of floral parts in classification has long since fallen away. The whole plant is used, leaves, roots and all. On the other hand, the tactile qualities of tendrils and climbers come to mind. The Keeper of Botany in my early days at the Natural History Museum was Robert Ross, who was known universally by the female staff as "octopus Ross." I mentioned earlier the small lift to the left of the main entrance that takes staff and visitors up to the Herbarium, and the Keeper's office which was then on the same floor. Women were warned never to go into the lift with Ross, or they would risk an attack of the tendrils. It was regarded as a kind of occupational hazard of working in the Botany Department. The women concerned seem to regard the memory of their encounters with amused resignation. Jenny Bryant recalls going into the lift with Ross while he was holding two large books in each hand; she still doesn't know how the wandering hand appeared between the first and second floors. For his part, "octopus Ross" never bore a grudge when his advances got nowhere—it seems he rather expected it, but had a go anyway. He was always immaculately besuited when I saw him, a short man and well groomed, the very image of a perfect gent. Of course, his behaviour would never be tolerated today. For some reason, it comes as something of a surprise to learn that his favourite pastime was morris dancing.

If Robert Ross' predilection was a common one, the story of Herbert Wernham was a little odder. He was a curator in the Botany Department. In the 1910s he wrote on the suitably named madder family, Rubiacea. Edmund Launert told me that Wernham was always broke. He frequently took his wages in cash. He was soon relieved of his earnings by two women who appeared every Friday on the dot at the back of the Museum. Even so, he always had something at the pawnbrokers. But the most extraordinary part of his story was that after he died a card index was found which contained a series of neat entries filed in alphabetical order. On each card was the name of one of his sexual conquests accompanied by a neatly pinned sprig of pubic hair. They might have been so many delicately coloured ferns. It seems that the instincts of the systematist were so deeply ingrained that he must perforce make an archive, duly arranged. Once a curator, always a curator.

Robert Ross was an expert on diatoms. Even in old age he could be consulted on matters of their nomenclatural minutiae. Diatoms are

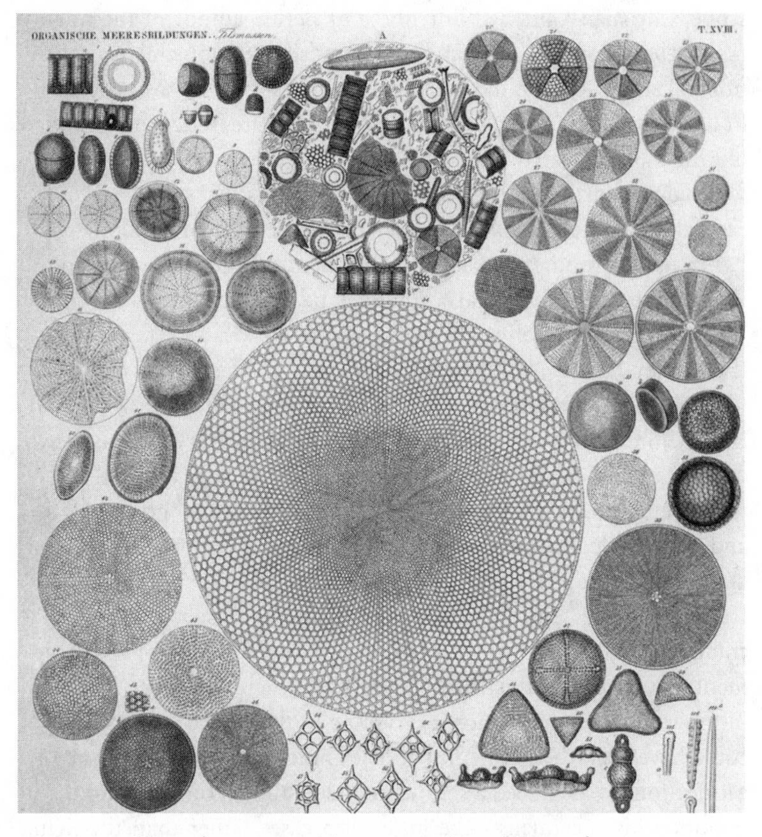

Nature's most exquisite constructions: diatoms from
Christian Ehrenberg's *Mikrogeologie* (1854)

minute single-celled plants that are ubiquitous in freshwater and marine environments. They are one of the most important components of the marine phytoplankton—and thus form much of the basis of the whole marine food chain. They are also very important bottom-dwelling organisms in lakes and waterways. Charles Darwin would have called them infusoria, which always reminds me of a tisane more than an organism. In his account of the journey of the *Beagle* (1839), he recorded the natives of Tierra del Fuego using a diatom-rich "earth" as face paint. The species were identified by that early hero of the microscope, Christian Gottfried Ehrenberg of Berlin, author of the mighty *Mikrogeologie* of 1854. Diatoms have exquisite skeletons formed of the mineral silica—silicon dioxide—that are composed of two valves that fit together like an old-fashioned pillbox. They are fretted and sculpted like the most delicately worked Islamic filigree.

I have a particular connection with diatoms that goes back to my early childhood in Ealing, the "queen of the suburbs" of London, where I grew up. My mother had a childhood friend, Catherine Morley Jones, whose parents lived in what would now be regarded as a rather grand Victorian house in Ealing. As a nine-year-old I was once taken to visit the Morley Joneses, and I recall a curiously old-fashioned household with ticking grandfather clocks and much dark-brown furniture. Mr. Morley Jones was a tall, distant and erudite figure. Everyone was in awe of him, and he was referred to as "father," as if he were a priest. But he treated children as if they were grown-ups, which was a novelty to me, and not unwelcome. He took me into his study and allowed me to look down a microscope for the first time in my life. He studied diatoms, and I remember the sense of amazement at the microscopic beauty that could be revealed just by using a strange piece of brass-and-glass equipment. It was no less of a surprise to learn that there was so much hidden in the world and that almost everything in the landscape was *alive.* It could have been the moment that determined much of my future. Mr. Morley Jones was indeed a serious amateur student of the diatoms. He mounted his specimens—the little silica cases rather than the living diatoms—on microscope slides labelled with his own tidy hand. Half a lifetime later I checked with the then expert on diatoms in the Natural History Museum, Barrie Paddock, and sure enough there was an entry for a substantial Morley Jones collection, donated to the Museum for the use of future students *pro bonum publicum.* Some of them must have been those same slides that I had looked at when I was still in short trousers, and which made me change my mind about being an engine driver. I experienced a curious frisson when I saw the old slides.

Diatoms are still an active field for research. They cross over into palaeontology, too, because their glassy skeletons have a good fossil record. David Williams is the current diatom guru, a man who carries the art of lugubrious gloom to humorous extremes. He has been work-

Lake Baikal, where diatoms, among many other organisms, have undergone a separate evolution

ing in Lake Baikal, the huge freshwater body in Russia, and a centre of endemic evolution. This study is in collaboration with University College, London, and the Limnological Institute of Irkutsk, and it is a good outcome of funding by the Darwin Initiative. The museum in London has put a lot of effort into projects under this heading. The Initiative was announced by John Major, then Prime Minister, at the Earth Summit in Rio in June 1992. It provides grants to help towards conservation of biodiversity, particularly if there is an element of sustainable development. It is intended to help developing nations, or areas where British expertise would be useful. Botanists from South-east Asia and South America have been trained up as taxonomists on Darwin money. I am sure the grand old man would have approved of this particular use being made of his name.

Lake Baikal is the world's oldest lake; in one form or another it has been around for about thirty million years. This accounts for much of its biological uniqueness, for there has been time enough for evolution to work its creative wonders. It is also the deepest lake anywhere—1,637 metres at most, so there is ample opportunity for organisms to become adapted to different depths. There are probably more than 1,500 species unique to it, nobody is sure yet of the final number. The lake holds 20 per cent of the world's freshwater, and, since water is likely to be the most important political issue in the twenty-first century, I suspect that the world at large will become much more familiar with Lake Baikal than it is now. David Williams and his colleagues have been scraping diatoms off the stones around the edge of the lake for several years—it gives David an excuse to be gently mocking about how he spent his summer: "scraping stones," he says mournfully. They have added many species of diatoms to the lake's biological inventory—there may be more than five hundred species, of which up to half may prove to be endemics. Like so many other organisms, diatoms also evolved in isolation in Baikal. But there are some diatoms, like *Eunotia,* that have species known elsewhere; these reveal some interesting patterns in biogeography. There proved to be a number of taxa in Baikal that were known widely around the Pacific Rim, even as far away as Vietnam, and the western United States. In some cases, however, these species were known only as fossils. They must have become extinct locally, but persisted in Lake Baikal. It seems probable that the distributions of the species we see today are related to ancient geographical distributions—ones that pertained even before the last ice age. The present species ranges are like a ghost of this former world. But evolution continued its work on the survivors to produce a series of species peculiar to Lake Baikal. The biological world never stands still: sometimes it is the smallest organisms that help to reveal the dynamics of our mutable biosphere. Few people even know that diatoms exist, yet the stories they have to tell may chart shifts of climate that are crucial to understanding how the world came to its present state. That is why it is so important to support experts in small and less showy organisms—they may be less glamorous than orchids, perhaps, but tiny diatoms may yet be informative of great matters.

I admire the botanists whose work I have described in this chapter. A different selection of people would have been just as admirable—but then, this book is my personal museum, and does not pretend to be comprehensive. I admire the way botanists respect more than two hundred years of taxonomic enterprise; I admire their commitment to improving botanical knowledge in countries less privileged than Europe or the United States. For every scientist I have mentioned there is a curator quietly working away on the herbarium. I think of Arthur Chater, a relentlessly modest man and prolific curator of specimens. As a young man he was coupled with Ted Hughes as one of the young writers published by Faber and Faber. He became an authority on coins, such as those minted in Aberystwyth under Charles I, of which he made the best collection. Some ten thousand of his many photographs of Welsh tombstones and other monuments are safely held in the Archives Network for Wales—and all carefully catalogued by Arthur Chater. It is impossible not to admire such versatility, such unfussy distinction.

I have to mention Dr. Norman Robson, too, if only to make amends. A few years ago I escorted Bill Bryson on a tour of the Natural History Museum for his book *A Short History of Nearly Everything.* Like millions of others, I enjoy Bill Bryson's humour and powers of observation: one of his books helped me to survive during a particularly depressing visit to Kazakhstan. During his visit to the Museum we met Norman Robson, and I explained that he had spent practically his whole working life on *Hypericum.* When the account appeared in Bryson's book, it seemed that Norman had spent his whole life studying a single species. In fact, the St. John's wort genus *Hypericum* comprises hundreds of species (460 at last count) spread around the world, including some species that are almost trees and others inconspicuous trailing herbs. Most have attractive star-like yellow flowers. Some species are of horticultural interest, others are poisonous to livestock, and others again seem to be as effective as conventional medicine in treating depression. They are currently under investigation by pharmaceutical companies. By any reckoning *Hypericum* is an important group of plants. Sorting out the many species of *Hypericum* is, truly, a lifetime's work. Norman is eighty years old, and has almost finished the task of completely revising the St. John's worts, down to the last small creeping weed, well after most people are beginning to find pruning the roses rather more than they can manage. He is a Scotsman of unwavering gentlemanliness. I have been feeling guilty about inadvertently giving the wrong impression of his worth ever since that visit, so let me state it clearly: the man is an unsung hero.

The relationship of the Natural History Museum Botany Department to Kew Gardens has been touched upon, but it should be added that Kew extends the virtues of the museum to embrace living collections, a place where vulnerable species can be coddled into survival. Plants, fortunately, can be rescued from extinction even if only a tiny population remains in the wild. The lust for rare orchids and their consequent over-collection in nature gives places like Kew Gardens an importance that outstrips their modest acres. It requires a marvellous skill on the part of the orchid man at Kew to raise these most difficult plants from their tiny seeds. The idea of a living ark is not one I wholly favour, lest it release the moral pressure on politicos and businessmen and developers to preserve natural habitats. Plants and animals belong under the sky, not in glasshouses and vivaria. But with this reservation: the efforts of these dedicated scientists to slow the extinction process that environmental degradation and climate change have brought upon the world must be applauded. The Millennium Seed Bank Project takes the whole conservation business a step further. By first drying and then storing them under cold conditions, seeds can be preserved for many years. The intention is to keep secure the seeds of something like 42,000 species by 2010. Kew and the Natural History Museum have pooled their taxonomic expertise for the good of the biosphere. Let us say the Project is an insurance policy against the destruction of habitats that a wise stewardship of the planet should never let happen. It is depressing that there probably will be no end to such destruction, for tropical forests and other habitats are still being destroyed despite the efforts of conservationists, sometimes at great personal cost, as in Brazil. It is a sad fact that to many people the loss of a plant species is of less moment than the loss of a football match. I hate the thought that the only record of a beautiful plant might yet be the grave of the herbarium sheet.

6

Multum in parvo

Martin Hall is far too modest a man to claim to have saved Africa from disaster. But that is what his knowledge of some of the less salubrious members of the Class Insecta has accomplished. He knows a lot about screw worms. These insects are members of the Order Diptera, which are so called because their wings have been reduced to two—whereas nearly all other insects have four wings arranged in pairs (*pteros* is Greek for wing, as in pterodactyl, *di* is two). The hind pair of "wings" has been transformed into specialized peg-like balancing organs called halteres. In common parlance dipterans are flies. Those of us with cats will at some time have had the experience of finding a dead bird in the house, under the bed perhaps, all heaving with maggots. These particular dipterans are nature's garbage disposal officers. We should, I suppose, be grateful to them for their useful work, but the more usual reaction is a convulsive shudder—maybe this derives from an instinctive response to warn us against eating bad meat. My father had fishing-tackle shops, and sold maggots to fishermen; they were known by the euphemism of "gentles" and came plain or coloured. "Half a pint of gentles—plain" was the commonest order. When I was helping in the shop I had to dig my hands down deep into the struggling mass. They were supplied to the trade by a farmer known simply as Wormy (he

Cochliomyia hominivorax—the New World screw worm fly. This adult was reared from a larva that came to the United Kingdom with a woman who had visited Trinidad and Tobago.

bred worms, too). I once went to his run-down farm on the western edge of London, in a bleak area that could not decide whether it was still countryside or just biding its time to become a cheap suburb. The worst moment was being shown the place where the gentles came from: a sealed room full of rotting carcasses, dripping profitable maggots. A scene like that would test the nerves of any budding naturalist. But it gets worse.

If feeding on decaying flesh is a good option, evolutionarily speaking, because flesh is nutritious stuff, it is only a small step to cut out the middleman—death. Feeding on *living* flesh is a logical progression in the dipteran way of doing things. Get in early, before the next fly. There are about two hundred species of flesh-eating flies. Martin Hall can identify their maggots within a couple of minutes, or so he tells me: to you or me maggots are just wriggling, pallid little tubes, but to Martin they are almost like old friends. A fly will lay its eggs in a vulnerable area—often a nice, moist place like around the lips, eyes or fundament. Then the larvae can hatch out and get to work. This habit may have come about by way of an intermediate stage whereby maggots consume bad and infected flesh around wounds—nasty but nourishing. These kinds of maggots have enjoyed a new lease of useful life recently, because they eat only rotten flesh and leave good muscle nearby untouched.* They even secrete some chemicals that encourage the formation of scar tissue. For those who have developed tolerance or allergies to antibiotics, a treatment by "maggot therapy" is being applied for the first time since the First World War. The maggots are sealed in a box fixed over the wound, and all the patient feels is a tickling sensation as the little hygienists get to work. When all the bad flesh has been consumed, the box is unstrapped and the contents disposed of. However, the next evolutionary stage—the exploitation of living flesh—is much less benign. Consider the Cameroonian tumbu fly (*Cordylobia anthropophaga*). The species name alone may furnish a clue. This unpleasant creature lays its eggs in places where it can smell the merest hint of urine. The larvae form "warbles" in the flesh of the victim in the most sensitive parts of the body. For some time humans were infected by way of eggs laid on the gussets of knickers hanging out to dry—providing direct delivery to the right kind of protected habitat. When the little beasts got to feeding, the pain and embarrassment can be readily imagined. Modern hot steam irons applied to the garments in the right place have helped to see off this intimate curse.

The screw worm takes the grisly story of being eaten alive a little further. The "screw" that gives the worm its name is actually a series of concentric spines. The maggots operate as a cluster: they secrete an extra digestive enzyme within the wound to smooth their progress into the flesh by turning it into soup. The barbs of the spines interlock so that the maggots cannot be dislodged by the rasping licks of an ox tongue. As the maggots move inwards they continue to breathe through their stubby back ends by means of spiracles—the openings to these "lungs" can be seen as tiny dark dots. The definitive screw worm is a New World fly, *Cochliomyia hominivorax* (the species name is the Latin

*The credit for putting the therapeutic uses of maggots on the medical map is often attributed to a Confederate officer, J. F. Zacharias, during the American Civil War. "During my service in the hospital at Danville Virginia, I first used maggots to remove the decayed tissue in hospital gangrene and with eminent satisfaction."

The feeding head of the larval screw worm *Cochliomyia*

equivalent of the Greek *anthropophaga* of the tumbu fly), which is particularly prevalent in Mexico and central America. It can progress from larva to pupa in under a week, and since an adult fly can travel many kilometres and lay thousands of eggs, these flies can quickly become a plague. They can reduce a cow to pulp. Even where the fly is native it is a scourge, but if it escaped into a place where animals were unprepared for it, the consequences could be disastrous. As its scientific name implies, the screw worm can feed on humans, too, in a comparable fashion. This is how it was first discovered. The French entomologist Charles Coquerel described the species in 1858 from a specimen taken from a man who died at Cayenne, the infamous penal colony in former French Guyana. It is in all our interests to stop the spread of this alarming creature, which is the point at which Martin Hall comes back into the tale.

In 1989 Martin Hall went to Libya and found screw worm larvae in the herd belonging to the Ministry of Agriculture. He told me it did not take him more than a few seconds to recognize the characteristic dark spiracles of *C. hominivorax.* The parasite had probably been introduced from South America in imported sheep. The Libyans did not like his diagnosis at all, and Martin was forced to raise some of the maggots in his hotel room to prove his point. He is not the type of man to let his expertise be challenged. Imagine if the screw worm were allowed to escape into Africa unchecked. Not only would domestic cattle suffer appallingly, but the fly would have wrought havoc on all the wild bovids across the continent, all those wildebeest and hartebeest and their relatives. The infestation could well have pushed rare species into extinction. It would have been an unprecedented catastrophe for an already blighted continent. In the Libyan context it was particularly galling that a common name for the parasite is the "American screw worm"—paranoid explanations of the infestation were bound to follow. Something had to be done to eliminate the pest, urgently. The technique used was sterilization. The idea was to release vast numbers of sterile male flies into the wild among the herds. If enough were liberated in this way, the chances of a female coming across a fertile male would be small; mating would lead to duff eggs at best; repeat the process several times and the viable breeding population would shrink to nothing. The crisis would be averted. The technique used involves sterilizing the males at a late pupal stage, by exposing them to a radioactive isotope of the element Caesium Cs^{135}. This is late enough in the development of the insect to produce a male normal in all other respects, including sex drive. The figures are extraordinary: sixty *million* pupae a week were air freighted from Mexico to Libya during the eradication programme. The logistics of producing so many maggots and pupae require that they are fed on massive quantities of a rather unpalatable meat pâté, a case of Wormy writ very large. Sterilization as a biological control was invented by Edward Knipling, whose name should feature with those of Edward Jenner and Louis Pasteur in the pantheon of those who rescued humans from misery. In 1995 was awarded the Japan Prize, which is almost the same as a Nobel, though less well known outside the scientific *cognoscenti.* As for the Libyan outbreak, the release of sterile males was continued until six months elapsed without capture of a fertile fly. With one bound, and billions of pupae later, the cattle were free. All this was achieved because an expert knew his maggots . . . The outcome is a plaque in the Food and Agriculture Organization of the United Nations, and the screw worm is now a "notifiable" pest at the Office of International Epizoology in Paris.

The price of freedom from flesh-eating flies is eternal vigilance. The kind of molecular evolution studies that we have already met in describing schistosomes confirm what many entomologists had suspected: that this nasty behaviour has arisen several times from different ancestral flies. The carnivorous flies in question appear on separate branches of the evolutionary tree, cousins under the skin. The Old World screw worm is called *Chrysomyia bezziana* ("Bezzi's fly"—and who would like such an animal named after him?) and learned its grisly trade independently of its New World counterpart. It is a native of sub-Saharan Africa and Asia. This species particularly infects open wounds, burrowing in using special hooks, and if such lesions are left untreated the parasite can cause permanent disfigurement. It is not averse to burrowing into soft fleshy folds elsewhere on the anatomy, rather like the gusset fly. *Chrysomyia* has arrived in the Middle East: in Iraq between the first and second Gulf Wars; in Bahrain, probably on shipments of sheep brought in to feed a growing population. Imported Arab horses are another possibility. Of the four horsemen of the apocalypse, only pestilence could be carried on the steed itself.

There are, of course, lesser parasites. The louse is an inconvenience rather than a death sentence. "Lousy" as an adjective does not rate as the absolute nadir; it's more of a generally disgruntled word, a crosspatch's rejoinder to a formulaic question about the state of his health.* The human louse has been having a lousy time of it in western countries since the 1950s; once ubiquitous, the head louse hangs on persistently in schools, and still sends a shudder through a mother who finds its "nits" (egg cases) on her children's hair. The louse was named scientifically, it will be unsurprising to learn, by Carolus Linnaeus himself: *Pediculus humanus.* There are two different subspecies, living on head and body

*Body lice do in fact carry several diseases, such as trench fever, which was a problem in the First World War. The Cambridge zoologist A. E. Shipley published a small book in 1915 entitled *The Minor Horrors of War,* dealing with the louse, bed-bug, flea and so on.

The plaque recording the part that taxonomy played in the eradication of the New World screw worm from Africa (*capitis* and *corporis*) respectively; the pubic louse ("crabs") *Phthirus pubis* is more distantly related. In nature there are some five hundred other species of lice living on many other hosts. One of the curators at the Natural History Museum responsible for lice was Bruce Frederic Cummings, who joined the staff in January 1912. He had—too grim for irony—a lousy time. He was dying from multiple sclerosis, what was then known as disseminated sclerosis. He was forced to resign from the Museum in 1917 because of ill health. But he recorded his struggle with decline in a remarkable memoir compiled from his diaries. His extraordinary account was published in March 1919 under the title *The Journal of a Disappointed Man.* Cummings wrote under the nom de plume of W. N. P. Barbellion. The initials W. N. P. stood for Wilhelm Nero Pilate, the author's selection of the most despicable people in history; Wilhelm was of course Kaiser Bill, this being the time of the First World War. Barbellion was a name emblazoned above a shop in South Kensington that Cummings passed every day. The *Journal* is a work of introspection so intense and unflinching that it leaves the reader exhausted. There is fury at his loss of hope, but since Barbellion was incapable of turning a dull phrase the text is also somewhat exhilarating. Of his wasting disease he writes: "Why this deliberate, slow-moving malignity? Perhaps it is a punishment for the impudence of my desires. I wanted everything so I get nothing . . . I am not offering up my life willingly—it is being taken from me piece by piece, while I watch the pilfering with lamentable eyes" (5 July 1917).

Several distinguished writers viewed his work with favour. H. G. Wells wrote the introduction to the *Journal*, and Marcel Proust and James Joyce were admirers. Other critics thought the work immoral and self-indulgent. The *Journal* is a hodge-podge of stories, health reports and observations on decline, which are riddled with guilt and illuminated by ecstatic flashes. It is impossible not to feel sympathy for this hypersensitive soul in perpetual torment. His time in the Natural History Museum was not in any sense a happy one. He was clearly devoted to zoology, and returns to this infatuation with the regularity of a mantra. He taught himself comparative anatomy from specimens borrowed from the Plymouth Marine Laboratory. He desperately wanted to be a professional natural historian. "In the repose of the spa-

Bruce Frederic Cummings, a.k.a. W. N. P. Barbellion, author of *The Journal of a Disappointed Man*

cious laboratory by the sea, or in the halls of some great Museum, life with its vulgar struggles, its hustle and obscenity, scarcely penetrates. Behind these doors life flows slowly, deeply" (4 March 1911). He wanted to know everything, to suck in the whole of nature, but was awestruck by the limitations of his memory and his persistence. He equally wanted to experience the life inside every pub and on every street corner—and women: he had an unquenchable voracity for experience, even as his health dictated a contraction into neurasthenia and introspection. He certainly did not wish to be any kind of functionary in the natural history world. "I don't want to be worrying my head over remedies for potato disease, or cures for fleas in fowls. Heaven preserve me from becoming a County Council lecturer or a Government Entomologist" (30 June 1911). The latter is exactly what he became. He yearned to be a great comparative anatomist, the Richard Owen of his day. Instead, he was destined to write pamphlets on lice, even as his serving contemporaries suffered from their unwelcome attentions in the trenches—an ordeal he was too unwell to share. He felt he was intended for nobler things. "I gave evidence as an *expert* [his italics] in Economic Entomology at the County Court in a case concerning damage to furniture by mites for which I am paid £8–8s fee and expense and travelled first class. What irony!" (8 October 1913). There is something of disdain for the commercial creature in his attitude. He wanted to fly nobly above the commonplace, to be admirable and remarkable and justly famous. He excoriated himself for failing to achieve his dreams, and yet how realistically he knew that they were just dreams. He was in many respects a thoroughly modern antihero. He wanted to know all of zoology, yet found that reading even one number of the German scientific journal *Zoologische Anzeger* cover to cover was beyond him. "Zoology alone was sufficient to baulk my puny endeavours," he wrote in 1913. "How hopeless it all seems!" He continued: "I shot up like a ball on a bagatelle board all steamy with zoology (my once beloved science) but at once rolled dead into the very low role of Economic Entomology! Curse . . . Why can't I either have a first-rate disease or be a first-rate zoologist?" It is that phase in parentheses—"once beloved" zoology—that saddens me. Everything, even the zoology he preferred to money or fame in his early diary entries, was subsumed under his existential gloom. The Museum was no cure; instead it became just another symptom.

A little further down the scale of parasites we have the flea. And the complete antithesis of poor Bruce Cummings was the world authority on fleas, Miriam Rothschild. Where Cummings was woefully insecure and met an untimely death, Miriam Rothschild had the confidence that comes with a famous name and a considerable fortune, and she lived to a great age. Her long life overlapped with that of Barbellion, and lasted into the age of biological molecular sequencing. The Rothschild family is one of those dynasties that almost everyone is aware of in a general way but whose details slip away untouched in a welter of discretion. There are three branches of which I am aware: the bankers, the wine makers and the natural historians. No doubt there are more twigs, and they certainly interlink in complex ways, since some of the naturalists were also able bankers. The bankers still own one of the last great private merchant banks in the City of London. This is a hermetic world that makes the back rooms of the Natural History Museum seem almost like public property. As the historian Niall Ferguson has explained, these

Rothschilds interceded in the big events in history in an ever so understated way: Who else would or could have financed the purchase of the Suez Canal for Queen Victoria? The wine branch of the family produces clarets so distinguished (and expensive) that wine buffs whisper their names as if they were religious incantations: Château Mouton Rothschild, Château Lafite . . . most of us dream that Uncle George will leave us a bottle or two in his will instead of the Maserati. The natural historical branch of the family was founded by Walter, second Baron Rothschild. He donated his private collection to the Natural History Museum on his death in 1937. This is the collection to be found in his old house at Tring, in Hertfordshire, thirty miles from London, where the ornithological collections of the Zoology Department now reside, eggs and skins in their thousands. The collection of domesticated dogs on display is another kind of "museum of a museum," a taxidermist's Mecca, a canine compendium. Walter's brother Charles was another dedicated natural historian who combined a successful business career with an intense interest in the taxonomy of fleas.

Miriam was Charles' daughter, and was devoted to him. When he left his collection of fleas to the Natural History Museum, she dedicated more than twenty years to its study. Her father had committed suicide in 1923, probably because of depression brought on by suffering from *encephalitis lethargica,* that mysterious "sleeping sickness" that followed upon the First World War. George Hoskins and Miriam Rothschild's great work *An Illustrated Catalogue of the Rothschild Collection of Fleas (Siphonaptera) in the British Museum (Natural History)* (five volumes, 1953–71) was the ultimate tribute of the talented daughter to the naturalist father. Miriam's natural habitat was looking down a microscope, and those who said she preferred animals to humans were probably not far from the truth. She was known as the Queen Bee. Her researches showed that the reproductive cycle of the rabbit flea was tied into that of the hormonal cycles of their host. This had relevance to stemming the plague of rabbits in England in the middle of the twentieth century by the introduction of myxomatosis. She became interested in the chemistry of the pheromones that induce sexual attraction in insects. She pioneered conservation techniques on her estate at Ashton Wold, near Peterborough—leading by example, and founding the notion of "think globally, act locally." She realized the importance of small areas

Walter Rothschild atop a giant tortoise

like roadside verges in maintaining biodiversity, something that is now adopted by many a Wildlife Trust. Miriam carried on working even as her sight failed, right up to the end. Her last scientific contribution postdated her. When I was a young employee at the Museum she was a Trustee, and the first woman ever in that role. We all looked forward to a glimpse of her. Her battered old Rolls-Royce would glide up the ramp at the front of the Museum, and the Queen Bee would step out in some rather large black dress—but sometimes still wearing the gumboots in which she had stomped around her paddocks that morning. As my mother used to say, she didn't give a damn. Her chauffeur was a thin man, with a pockmarked complexion, decked out in a reddish livery. He used to stand around scratching himself in a bored kind of way as if one of the samples had escaped.

I ought to explain some more about specimens at this point. When I went on my pioneering wanderings through the bowels of the Museum, and finally reached the entomologists, the drawers that I opened were full of pinned insects. A pin impaled each specimen through the thorax, like the stake that finished off Dracula in the many films starring Christopher Lee. This is how you fill up a collection of butterflies and beetles: you pin them down. Dozens or even hundreds can fit in a drawer. But the tiny insects we have been discussing are obviously too small for such impaling. They need a different method to preserve them for posterity. Some are preserved in spirit in small phials as whole specimens. But most of them are "prepared"—spread out on microscope slides, often dissected into several pieces, with their cuticles chemically cleared. Depending on the type of insect concerned, there may be wings laid out, or the "hairs on legs," or the genitalia, or the mouthparts. It is a way of defying time, of turning a tiny, living thing into an archive of life. The process somehow recalls the twilight lines of T. S. Eliot in "The Love Song of J. Alfred Prufrock": "When the evening is spread out against the sky / Like a patient etherized upon a table." It is a moment of existence anaesthetized or protected from decay. Aphids (greenfly) have their soft and squashy innards sucked out, and their thin outer coverings preserved, like a coat that perfectly models the body. Then a truly vast number of specimens can be stored inside a single cabinet. Parasitic wasps are minute creatures that lay their tiny eggs upon larvae of other insects, which then hatch to consume their hosts. They, too, are preserved dissected in ranks of slides, all neatly labelled at one end. Small creatures somehow encourage neatness in curation.

Insect pests often do come in small sizes: think of weevils in flour, furniture beetles with their tiny "woodworm" holes, greenfly on the roses, clothes moths in bottom drawers. A thing of small dimensions creeps in through cracks and does damage: it is "The invisible worm / That flies in the night / In the howling storm" that sickened William Blake's rose. But the tables can be turned: some diminutive insects can be employed to positive ends. Parasitic wasps can be used to target particular pests, because they are very choosy about their victims. Natural History Museum entomologist Andrew Polaszek explained to me how this form of biological control has been used successfully to combat the depredations of certain white flies. Despite their name, these tiny flyers are not flies—they are actually miniature relatives of the aphids. They also resemble minute white moths when they infest

Punch's view of the obsessive lepidopteran collector, 1985

my cabbages—though of course they aren't moths either. But they *are* virus vectors: one species has been shown to carry a hundred species of virus. This is not good news for any plant that the white fly chooses to feed on. By the kind of paradoxical nomenclatural wheeze of which nature is apparently so fond, there is a *black* white fly called *Aleurocanthus woglumi,* which has been causing much damage to the orange and grapefruit crops in Trinidad. Like so many pests, this species is a stranger which has got out of control: its natural habitat is somewhere in Asia, in China or India. It lacks natural enemies outside its home territory, so it breeds without restraint on the citrus plantations of the New World. It had already made its presence known in Florida, a state that is a vast producer of orange juice. Help came in the form of a minute species of parasitic wasp (*Encarsia perplexa,* colour plate 11) specialized to consume this white fly species alone: like the *Alien* in the film of the same name, this wasp flourishes and grows inside its host's larva, consuming it alive from within. It pops out in due course in place of the host that should have hatched all ready to infect yet more white flies. Vast numbers of the little wasp were bred up at the University of Florida at Gainsville and released on to the stricken plantations. The scheme worked well: the fear that the wasp might turn its attention to some other, wholly innocent species was not realized.

But if the removal of orange juice from breakfast menus might have been a source of irritation in the households of the west, it would be as nothing compared with the consequences of removing cassava from the diet of developing countries. Cassava (*Manihot esculenta*) is one of the most efficient of all domesticated plants for manufacturing carbohydrates—which is a good working definition of a staple foodstuff. What the potato was once for the Irish, cassava is now for the tropical peasant. Originally a South American species, it is the most important food crop in sub-Saharan Africa. But the crop was viciously attacked by a mealy bug from its home territory called *Phenacoccus manihoti;* these plant bugs look like scabs of cotton wool at first sight, but underneath the fluff they suck out all the goodness from their hosts till the plants turn yellowish, then sicken and die. A tiny parasitic encyrtid wasp from Paraguay came to the rescue; it was named *Anagyrus lopezi.* Mealy bugs of the cassava-consuming species were its favourite food. It does not figure among the most heroic creatures on the zoological hit parade, but this little creature probably saved many thousands of lives. Its recognition was the product of expertise in taxonomy, extensive knowledge of "hairs on legs," a skill that might be considered esoteric until it becomes vital.

I have speculated from time to time as to whether researchers come progressively to resemble the organisms upon which they research. There is a certain amount of evidence for my thesis, although I doubt whether it would survive rigorous statistical examination. For example, I have come to resemble a trilobite as I have got older, particularly as regards the middle lobe of my anatomy. A former Keeper of Zoology, Colin Curds, worked upon very small organisms that lived in sewage, and he did indeed resemble some kind of obscure micro-organism such as one might observe on a slide, being a little man with a pointy beard who darted around in a manner rather like *Paramecium.* Gordon Corbet, who worked on small mammals, was a Scotsman with a hesitant manner and a nervous way of speaking; for some reason he reminded me of a vole—the way these animals pause momentarily, whiskers twitching. But there were some particularly striking examples in the Entomology Department. W. N. P. Barbellion (diary entry, 20 April 1914) had noted such resemblances when he remarked: "an Entomologist is a large hairy man with eyebrows like antennae." Ian Yarrow worked on bumblebees (particularly the genus *Bombus*), and he was a man with a very comfortably rounded middle. He had a fondness for furry jumpers, which gave him a thoracic look, and on one occasion I saw him wearing a woolly sweater with broad horizontal stripes. Furthermore, he used to hum to himself: once I followed him down the front steps of the Museum as he bounced from foot to foot going "bzz, bzz, bzz" quietly. At the risk of spoiling my thesis, however, I should add that he smoked a pipe, which is most unbee-like. Dick Vane-Wright had one of the longest possible Museum careers, and rose slowly through the ranks to become Keeper of Entomology at the end of the twentieth century. He works on butterflies, particularly showy tropical butterflies, and, when compared with most entomologists, has an appropriately flamboyant personality; he was briefly fond of outrageous waistcoats with paisley patterns. He was also very partial to attractive women and took a genuine delight in beauty, and I cannot help thinking of him flitting butterfly-like from flower to female flower. Somehow, even his propensity for playing the trumpet seems concordant with his being a lepidopterist: all those long tongues pushing out of the mouth. If reincarnation were a possibility I would imagine Dick Vane-Wright as a red admiral butterfly opening his splendid wings in the sunshine upon a *Buddleia* bush. On the other hand, Barry Bolton, the incomparable ant man, was always enormously and single-mindedly industrious, just like . . . well, I hardly need to labour the point. He could also release a blast of acid when the occasion demanded it. Several of the beetle men developed hard carapaces and tended to avoid the daylight. The colourful botanist Edmund Launert has been mentioned—and he has published a standard work on orchids—while the fern man Clive Jermy tended to shun the limelight, and worked away modestly hidden in the shady crannies of the Cryptogamic Herbarium. As for the former spider man, he had eight legs and spent all day on a web (all right, I made

Dick Vane-Wright dressed as his subject of study: the Lepidoptera

that one up). As final proof, I am told that recently I have become more of an old fossil than ever before. I rest my case.

There are some corners of the Entomology Department where few people ever venture. The towers of the Museum are the farthest reaches of Gormenghast. Even to find one's way to the towers is an exercise in map reading. The visitor has to go through one door after another apparently leading nowhere. Then there are thin flights of steep stairs that go upwards from floor to floor; I am reminded of a medieval keep, where one floor was for feasting and the next one for brewing up boiling oil. I discovered part of one tower that could be accessed only by a ladder stretched over a roof. Nowadays, the towers do not have permanent staff housed there—something to do with them lacking fire escapes and not complying with some detail of the Health and Safety at Work Act of 1974. Instead there are empty rooms, or ones holding stacks of neglected stuff. A hermit could hide here, undiscovered. From the top of the towers there is a wonderful view; London is laid out just "like a patient etherized upon a table," new landmark towers reaching above the fine old heights of St. Paul's Cathedral in the distance to the east. But then: in plastic bags laid out on a bench there are dead piglets being eaten by maggots. On a side bench there is a register labelled "Calliphorid culture book." This is the room where experimental decomposition of flesh by blowflies is under way under the enthusiastic supervision of Amoret Whitaker. Piglets are rather more available for experiment than human corpses. When the maggots have done their work, the piglet is reduced to skin and bones. Amoret waved such a filleted animal at me; it looked like a grisly deflated copy of the original rendered in parchment. Another bag was full of pupae—the next stage in the life cycle of the flies, which mostly belong to the Family Calliphoridae. These experiments are in the service of forensic entomology, an area of science where the police and the Natural History Museum enjoy a brief liaison over a body. Like undertakers, forensic entomologists seem to be a habitually cheerful lot. Ken Smith, Amoret's predecessor, was quite happy to be known as Lord of the Flies. He published the standard textbook on forensic entomology in 1986. At that time the scientists gave weekly lectures to one another as part of the activities of a Scientific Officers' Association. Ken Smith's talk featured projection slides of decaying bits of anatomy. More sensitive members of the audience soon started to look green and staggered towards the door. As the slides got more graphic, more people left. Only a few policemen were left at the end, taking notes.

The entomology of corpse decay has become quite a sophisticated science. The premise is quite simple, at least as far as determining the time of death is concerned. There is a succession of different insect species with different life cycles that strip down a body. When a body is discovered, it is possible to determine approximately how long it has been lying around by capturing the insects and/or their larvae, and counting and identifying them. By far the best indicators are the common or garden bluebottles and their relatives. They get in first, and their life cycles are short and known in detail. No sooner does a corpse get laid out in the woods than it exudes a delicious odour—at least to calliphorid flies like bluebottles. They swarm from afar. Eggs are laid, maggots hatch out, consume the flesh, then pupate for some days—and subsequently hatch out again all hungry and ready to mate. So by examining the moult stages of the larvae if they are still alive, or seeing whether there are empty pupal cases around the corpse, it should be possible to tell how long the body has been exposed to nature. In temperate climates the rate of the fly's life cycle will vary according to the time of year—the warmer it is, the faster it goes. Then there are many other insects that take the process of decay further, and all of them have wonderfully suggestive names: the coffin fly (*Conicera tibialis*); the burying beetle (*Nicrophorus* species); the sexton beetle (*N. orbicollis*); and the larder beetle (*Dermestes*), "reported as reducing a human body to a skeleton in only 24 days." This last is the beetle still employed at the Natural History Museum to clean flesh off the bones of specimens destined for the osteological collections. Beetle larvae can cope with the tougher bits that blowflies eschew, such as skin and ligaments and tendons. They will be the last ones to leave our bones alone when our time comes. Put a late fellow of infinite jest into a pit with *Dermestes* and in no time you have the skull of a Yorick. The tank in which this process is allowed to happen in the cause of science is known as a Dermestarium.

There are many additional insect species that will call in on a decomposing corpse for a little bit of protein on the side, so a forensic entomologist must know a lot about the Class Insecta in toto. Zakaria Erzinclioglu, known universally as "Dr. Zak," former Director of the Forensic Science Research Centre at Durham University, has described with Sherlock Holmes–like precision in *Maggots, Murder and Men* the inferences involved in some of his cases. The discovery of an obscure species of winter gnat and the absence of bluebottles might go to prove that a murder happened in the cold part of the year. You can almost hear the pipe being tapped out against the fender. "Surely you are cognizant of the habits of the winter gnat, my dear Watson." There are many subtle variations in the way entomological evidence can be used, given differences in site and time of year and the covering, or not, of bodies—and whether a particular species of beetle happened to be in the vicinity at the time a cadaver was deposited. A famous case brought to the Museum involved the examination of "sweepings" from a murder site. Apparently the suspect claimed that carpets had been removed due to an infestation of fleas (suspicious stains, naturally enough, providing the alternative explanation). Microscopic examination of dust from the site showed none of the evidence of larval or adult fleas that should have been present if the proffered explanation were true. Case closed. This is negative evidence, of course, but perhaps not dissimilar to Holmes' famous case of the dog that failed to bark in the night.

One place where the forensic entomologist *can* get to work on

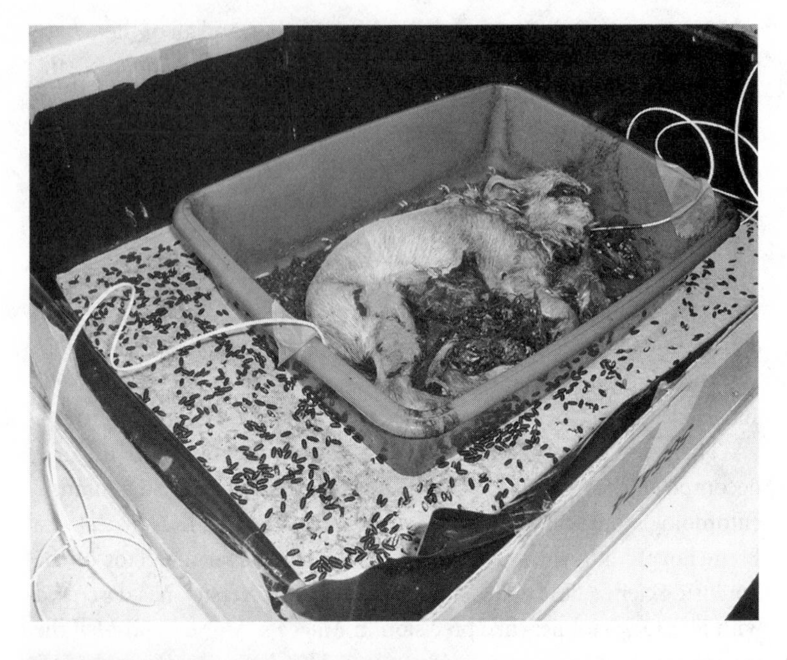

Experimental decomposition of a piglet by flies (*above*) and, left, the chief agent of rapid consumption of flesh, the bluebottle *Calliphora vicina* (with larvae)

human bodies is the Department of Anthropology at the University of Tennessee, Knoxville. This utterly respectable academic establishment has become the centre of a modern human bone collection. This may seem an odd thing to aspire to, but it is actually rather rare to have a collection of skeletons for which the name, history and biography are known. This growing collection will provide a statistical basis for future anatomical and pathological studies; I am told that 75 per cent of the bodies used are donated by the families of the departed. This also provides an opportunity for the forensic entomologists to study decomposition under observed conditions, for the meat must be removed from the cadavers before they are incorporated into the collection. The bodies are left out in the woods to let nature do the work: those in the trade refer to it as the "Body Farm." I must say that you have to be a forensic entomologist to contemplate with dispassion the fly-blown corpses scattered around in a progressive array of decay. Most of us would probably prefer to study butterflies.

But then, butterflies and moths are also the insects that inspire the acquisitive collector's obsession. Small flies and aphids are strictly for the dedicated scientist. There are collectors who simply *have* to own a specimen of some beautiful, rare or famous species of Lepidoptera. A desperate satisfaction is to be had from being the owner, the one who gloats over possession in the secrecy of a private study, the hoarder. John Fowles' first novel, *The Collector* (1963), delineates the character exactly, the "hero" of the novel graduating from capturing butterflies to kidnapping a beautiful girl. Fowles gives his collector the name of Clegg; I don't believe that it is a coincidence that cleg is the country name for an altogether unattractive insect—the persistent horsefly that will not leave you alone until it has tapped into your blood. That most fastidious stylist among novelists, Vladimir Nabokov, had a parallel life as a lepidopterist. His speciality was the blue butterflies, a group that manages to be both complex to understand and delicately beautiful, so I suppose that is entirely appropriate. Near where I live small blue butterflies flutter like so many animated harebells on the chalky downs: they don't want to be pinned down. Nabokov published extensively on these butterflies, and I estimate that his readership for such scientific papers might be one-thousandth that of the novels. But the famous writer regarded his entomology as almost as important as his fiction, which provides an encouraging change from those who measure distinction by the yardstick of fame alone. The genus *Nabokovia* is named for him.

Those who work professionally with butterflies soon lose the desire to own their own collection. If one can look through endless drawers of perfectly pinned specimens whenever one wishes, ownership of them quickly begins to seems irrelevant. I lost my desire to own trilobites when I could look at the national collections as often as I wanted. However, there have been butterfly collectors who have not been above pillaging the ranks of Museum specimens for their own secret stashes. Dick Vane-Wright told me about Colin Wyatt, a well-known collector in the 1940s. He was quite famous as a mountaineer and adventurer, and made a good living as a circuit lecturer. He appeared regularly in the Museum to study Apollo butterflies, as a respected visitor, arriving with his briefcase holding sandwiches and notebooks. Eventually, an observant curator realized that specimens seemed to have been disappearing after his visits, and a covert watch was kept. It transpired that he had installed a false bottom in his case: he was a real kleptomaniac. The bottom of the case could be lifted out, and hidden beneath it there was a cavity lined with cork into which butterflies could be pinned. Dozens of specimens could be secreted away in this fashion, the work of a few seconds. He was successfully prosecuted in the end—but only after he had changed the labels on hundreds of specimens to favour himself.

In writing about the hidden Museum it is tempting to concentrate on the "parasites" and the "butterflies," those people and case histories that induce a shudder or a smile. I should make more of the unsung heroes of the Entomology Department: those who labour away for their whole lifetimes to elucidate a chunk of the insect world. Because there are just so many insect species, these labours are bound to be Herculean. The qualities needed for such a systematist are preternatural persistence, tireless organization, a prodigious memory, an eye for detail, a capacity to draw, a talent for writing clearly, a will for finishing things and an ability to snarl at intruders. The ability to discourage unwanted distractions is vital to the completion of the Great Work; those lesser souls who always like to chew the fat and put off completing that scientific paper until the day after tomorrow will never quite get there. If genius is the infinite capacity for taking pains, as famously defined by Thomas Carlyle, then these people are truly geniuses—and possibly saints at the same time, a most unusual combination in my experience. Barry Bolton spent his working life on ants until he retired.

A smallish man with white hair and a staccato manner, he reminded me of William Hartnell, the first Doctor Who. His devotion to ants excluded every diversion, and all Museum management exercises were ignored or repelled, sometimes with an ungracious expletive. Ants demand microscope work, hours of it, and skills in dissection, and a capacity to remember all those other ants. The end result of all that labour is the completion of the most comprehensive systematic treatment of ants in a series of massive publications. Barry considered every known ant genus in his 1995 book *A New General Catalogue of the Ants of the World* (all 504 pages of it). One of the greatest of all living evolutionary biologists, E. O. Wilson of Harvard University—also a specialist on ants—confessed to being slightly in awe of Barry Bolton. As this is written, a new and even more comprehensive work dealing with all 14,550 species and subspecies of ants is being advertised: *Bolton's Catalogue of Ants of the World.* Other authors are involved with this work, but it seems that Barry has now achieved the immortality of Gray, of *Gray's Anatomy.* Anatomy *is* Gray, just as ants *are* Bolton.

One of Bolton's predecessors, Peter Mattingly, was one of the leading researchers of his time on mosquitoes. Such was his fame in the early 1980s that when a visitor arrived from Africa to see Dr. Mattingly—only to find that he was on leave—he still requested to be shown into his office. There he spent several minutes in silence gazing at the empty chair normally occupied by the guru. He just wanted to absorb the ineffable vibrations. But within the Natural History Museum Mattingly was famous for being the most absent-minded of all scientific staff, and this in a profession where absent-mindedness is regarded as part of the job description. He was quite bent over, particularly at the top of his back—not I believe because of any congenital deformity, but as a consequence of many years peering down a microscope. The condition is known as "microscopist's hump." So when he walked about the place it was with his head leading the way, glasses perched somewhere near the tip of his nose. When he left work in the evening he took off down the front steps of the Museum at a terrific lick and marched straight across the Cromwell Road, briefcase dangling from one hand. The Cromwell Road is one of the busiest thoroughfares in London, so how Mattingly managed to escape being run over was a mystery. Possibly thinking about mosquitoes all the time protected him, or maybe he actually had an insect-like sensitivity to approaching hazards and possessed the speed of reaction of a housefly. Mattingly collected his keys every day from the key pound. On one occasion in July he arrived in the usual way and wordlessly scooped up his keys, moving head-first towards his office. A few minutes later a woman ran in, breathless. "Has Dr. Mattingly arrived?" she gasped. When the answer was given in the affirmative, she explained that she was Mrs. Mattingly and that all the family had been waiting in the car to go on summer holiday . . . but before he could be stopped Dr. Mattingly had grabbed his briefcase and shot off to the commuter train and come into London to work on his mosquitoes. He had forgotten all about the holiday. One of his entomological contemporaries told me of a time when the lavatory arrangements in the department were renovated, the upshot being that the Gents' toilet was replaced by a Ladies' on a nearby site. After several months, Mattingly happened to remark that they had rather mysteriously removed the urinals from the Gents' lavatories, and why would that be?

One or two people have toppled over into clinical madness. Peter Lawrence worked on the tiniest of insects, the springtails (Order Collembola). These wingless little creatures are often assumed to be the most primitive living members of the Class Insecta, and they certainly have a fossil history going back to the Devonian, about four hundred million years ago. If damp vegetation by a pond is turned over with the fingers, the chances are that some jumping little flecks will be disturbed, and that these will be springtails: they have a little "spring" on their body that helps them to jump. With around ten thousand known species, springtails provide plenty to occupy a working life, and Peter Lawrence knows a great deal about these little animals. He became obsessed with the idea that they were crustaceans rather than insects. He also suffered from what is now known as bipolar disorder. In the Museum, the condition had an opportunity to develop unchecked to more and more impossible extremes, while Peter was tucked away in his cubicle, until in the 1980s he was finally hospitalized. This might seem like a tragedy, but Peter would probably disagree. He is an extraordinarily creative person. In 1989 he published an account of famous "manics," *Impressive Depressives,* showing how such people during their "highs"

An example of the tiny, primitive wingless insect the springtail, much enlarged

achieved exceptional things—he would probably like the condition to be considered as much a gift as a curse. I received extraordinary letters from him during the 1990s. He sent wonderful, crazy collages—a mixture of drawings and cut-outs from magazines and written messages. I think they may have been works of art. I learned recently from Edmund Launert that Peter Lawrence amassed a collection of thirty thousand picture postcards, probably the largest of its kind in Britain—he supplied three-quarters of the images for an exhibition of such memorabilia at the Victoria and Albert Museum. The fact that he always wore clothes picked up from jumble sales, or that he collected pigeon dung from the towers of Gormenghast to supply his houseplants, seems rather unimportant compared with his idiosyncratic achievements. And since recent molecular work has shown that the crustaceans are closer to the insects than was thought twenty years ago, maybe the springtails still have new and interesting things to reveal about their ancestry.

There is no question that the Natural History Museum did attract oddballs, some of whose conditions were less identifiable than that of

Peter Lawrence. One or two were rich enough to require no financial support, and presented specimens, so the Museum became a kind of asylum for them where a life could be spent painlessly away from the real world. The Baron de Worms sounds like he should have been the authority on annelids, but he was an aristocratic entomologist of private means who was a perpetual presence in the Entomology Department in the middle of the twentieth century. He wore thick pebble glasses and a suit like a banker and was shaped very much like Humpty Dumpty in the famous Tenniel drawing from *Alice Through the Looking Glass.* He sported a dark moustache, and had very black hair that never went grey. I am told that he was frequently to be found waiting outside the Ladies' lavatory, but I have been unable to establish exactly to what purpose. He attended all the meetings at the Royal Entomological Society, whose premises lay opposite the Museum at its western end, at 41 Queens Gate. Despite the fact that in those days the seats were very uncomfortable wooden pews, the Baron seemed to be able to fall asleep during the lectures, and punctuated the proceedings with loud grunts. If he liked the lecture, the grunts were different ones than if he disapproved. He was a famous glutton. A member of staff called Graham Howarth once asked him to dinner, and he established that he must come early and leave quickly. After consuming a three-course dinner, on his way out he asked his host if there were any decent restaurants in the vicinity as he was feeling a little peckish. And at one Entomology Department Christmas party some young wags put those expanded polystyrene flakes that are used in packing into one of the snack bowls—the Baron de Worms got halfway through a bowl before he realized anything was amiss.

Every so often, albeit rarely, major changes in ideas about insect ancestry do happen. Paul Eggleton demolished an order of insects in 2007. Termites are the voracious wood consumers of the tropics; they are also social insects unrelated to Barry Bolton's ants, but every bit as sophisticated. They have distinctive castes within each species designed to do different tasks, and all governed by hormones released from a gigantic, motionless queen lodged deep in the nest, who might lay thirty thousand eggs with no trouble at all. There are something like three thousand species of termites, of which just a few are injurious to mankind. Those that do harm us are serious pests: in the southern United States, termite damage is alleged to cost more than flood and fire damage put together. Their red tower-like mounds in the outback of Australia are among the most wonderful constructions in nature—orientated precisely to maximize the sun's beneficial effects, and "air conditioned" internally so as to maintain a temperature of 29.5 degrees centigrade. It has been one of my life's unfulfilled ambitions to eat the giant mushroom *Termitomyces* that grows deep inside the termite mounds in Africa—it is reputed to be delicious. For all my working life I have been accustomed to referring to the termites as belonging to the Order Isoptera—meaning "same wings," a nod to the similarity of their fore and hind wings. Paul's latest work based on sequence analyses of five genes spanning a number of termite genera and a series of cockroaches proves that the termites are classified with one particular kind of wood-eating cockroach—or "woodroach," *Cryptocercus.* Termites are thus likely to be highly evolved cockroaches—or, to put it another way, these hitherto separate groups of insects share a common ancestor.

The systematic result is that the old order must be revised in the most literal sense: termites become social cockroaches and must be classified together with these large insects in the cockroach Order Blattodea. Goodbye Isoptera. This proves that the hierarchy of a classification is always just provisional, and dependent on the state of knowledge. Even an Order can vanish as evolutionary history is better understood. There had actually been hints of the close relationship between *Cryptocercus* and termites before the new evidence came along: they share some very distinctive tiny organisms in their gut "flora." Science is often like this: an idea has been around for a while before new evidence suddenly pushes it forwards. And then researchers tend to think: maybe this example is not so surprising after all. The cockroaches are persistence incarnate; they have been around since the Carboniferous period, and will doubtless outlast mammals, and indeed just about everything else on our vulnerable planet. If there were a nuclear disaster of universal proportions, or global warming cooked us all, I can imagine cockroaches crawling out of cracks in the earth in the aftermath to clean up the mess. Nobody likes them much, but they are probably going to survive our passing. And if vegetation also survives in any form in the future, their relatives the termites will be there, too, demonstrating the strength of numbers and a prolific society—mindless but almost indestructible.

Some insects are particularly useful in monitoring the climatic changes that may yet secure the dominion of the cockroach. Steve Brookes works on chironomid midges. If you have seen a vaguely irritating black fuzz of tiny insects around the edge of a pond, the chances are that it will be composed of a mass of chironomids. Unlike the biting midges that drive one mad in Scotland in the summer, chironomids don't need to take a nip of flesh to reach maturity. Their larvae feed in ponds on small organisms like diatoms described previously. As they grow, they moult. Their cast headshields are readily preserved in the soft sediments that accumulate in the bottom of freshwater lakes, becoming effectively tiny fossils. Since different chironomid species live in different climatic regimes, they change in harmony with changes in climate: their little fossils can provide a good proxy for fluctuations in temperature. Because they are so small, they can be recovered in significant numbers from the tiniest sediment samples. Where deposition of sediment in lakes has been continuous over thousands of years, the chironomid fossils give us an historical narrative of climate change. This can be matched to independent evidence derived from other continuous historical narratives, such as those provided by the Greenland ice cores, where variations in oxygen isotope ratios provide a measure of past temperatures, or the sediments of the deep sea, where the tiny shells of fossil foraminiferans perform the same function. Naturally, a standard is needed to calibrate a given fossil assemblage in terms of the ambient temperature of the time. Steve Brookes and his colleagues made a profile of chironomid species' abundance in July running from the Arctic island of Spitsbergen, 80 degrees north, to southern Norway, 58 degrees north, and ranging in altitude from sea level to 1,600 metres. With the optimum requirements of any species so determined, these scientists were in a position to interpret cores through ancient sediments. The sampling can be down to a millimetre or two, representing time slicing of the order of a decade, applied over the last fifteen thousand years or so. There proved to be a large number of lakes that provided appropriate core samples for analysis. Thanks to their new

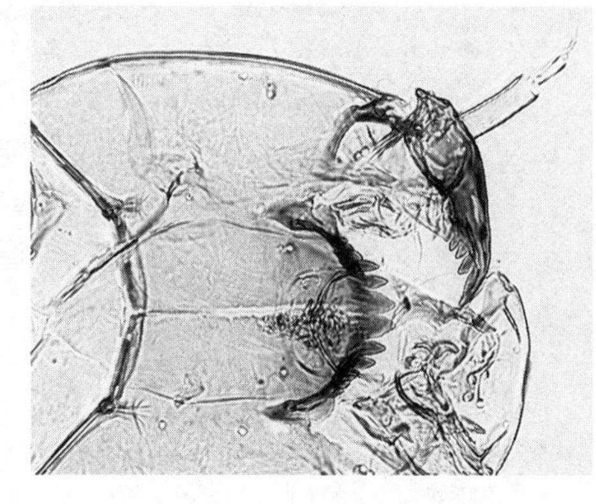

A microscopic slide of a midge head (*Heterotrissocladius grimshawi*)—an example of a chironomid used for studies of climate change

methods, the oscillations of the climate through warming or cooling could be monitored in the shifts of populations of the tiny midge fossils from the sedimentary samples; as the surrounding environment fluctuated, so did the dominant species of these tiny creatures. Sometimes the small size of insects is their greatest virtue. Mammoths and cave bears cannot be used with such precision. And now with the attention of the world focussed on climate change, understanding what has happened in prehistory has never been more important, in our attempts to sift out man-made from natural climate fluctuations. Study of the smallest things can sometimes attack the biggest questions.

When celebrating the sheer exuberance of the insects the beetles always come to mind. I have mentioned that estimating the number of insect species in the world—most of them still unnamed—has been the cause of much disagreement. The best I can do is to offer the opinion of the International Entomological Congress held in Brazil in 2000, when a symposium on the subject finished up with a questionnaire: the majority of entomologists believe that there may be something like eight to fifteen million species on our planet. There are an estimated twenty-eight million insect *specimens* in the Natural History Museum, including about a quarter of a million type specimens. The largest component of these vast collections is the beetles, which comprise something like 30–40 per cent of the insects. Peering into a drawer of pinned beetles is like looking down at an immense miniature army, all shinily

Above: A parade of beetles: just a sample of the most biodiverse group of organisms
Left: One of the largest and most distinctive beetles: the rhinoceros beetle

decked out in their battle gear. The footsoldiers are black and small, seemingly endless battalions of them, but here is a rank of larger iridescent green officers, and there a battery of rhinoceros beetles equipped with pikes, lances and halberds. A great number of beetle species remains to be discovered—no two coleopterists agree exactly on the figure. The 430,000 or so named species seem to be quite enough to be getting along with. Beetle specialists tend to be very specialized. They are likely to study just one family (or even a genus), not unreasonably saying that this is already a lifetime's work. The world belongs to beetles, which can live anywhere and eat anything. The key innovation during their evolution was turning one of the usual pairs of insect wings into hard covers, or elytra, which are used to cover up the other, flying pair. They can wrap their wings away until needed. Beetles get under logs and down into the soil and into dung balls and all kinds of tough places because, as my coleopterist colleague Peter Hammond puts it, they are "crunchy." I prefer to stick with my military analogy and think of them as armour plated. Anyone who has squeezed a ladybird and seen it walk away a few seconds later quite unfazed will know what I mean. And then the wings can be unwrapped and, delicate as a thistle seed, the little fellow flies away. They can feed happily on wood or pollen or mushrooms or oil spills, or whatever, and then fly to meet their mates when they need to. Some of the weevils don't even need water—they can make it from the wood or grain they eat.

I might say that the only resemblance between Peter Hammond and his objects of study is his beetling brow. Otherwise, he is classically handsome, smokes a pipe, and has the splendid chisel jaw of a leading man in a 1950s detective drama. He is devoted to a family of beetles called staphylinids, better known as rove beetles. They are active, flexible beetles with very short elytra—under which the other pair of wings is folded. If you turn over a rotten log, the chances are a dark-coloured "staph" will run away from the light in a few seconds. This family alone has forty thousand described species, and Peter thinks that there may be half a million different kinds of them. The species are often told apart by details of the genitalia, as well as by more obvious things like colour and ornament; Peter's colleague Jim Bacchus referred to monographs identifying these beetles as "prickture books." A pun might make the task of getting to know tens of thousands of unnamed species a little less intimidating.

I wonder how entomologists cope with facing this Mount Everest of unnamed species. Some tackle the mountain by the north face, head on, by trying to describe and name as many species as a life allows. An American entomologist called Alexander may hold the record; he worked on the crane flies, Tipulidae, and is said to have named some twenty thousand species, using millions of words to systematize their endless variety. Noble though these endeavours are, they are doomed to failure; there is just too much to do. In 1971–72 four young men from the Museum—Mick Day, David Hollis, Dick Vane-Wright and Peter Hammond—set out on a major expedition to South-west Africa, to collect as much insect material as they could in five months. They adapted an army truck as a field laboratory. They managed to purchase the old truck for the mighty sum of £300; then they grafted an army fire engine on to the front so that there was room for them all in the "cab." Dick Vane-Wright proved to be as good a carpenter as he was a lepidopterist and fitted out the laboratory in fine style; Mick Day turned out to be that most important person on any expedition, a talented field mechanic. Peter Hammond says he wasn't good at anything so he acted as a porter. The whole expedition cost a mere £3,000, including the vehicle, a sum that now seems laughable even allowing for inflation. During the trek across Namibia, Botswana and southern Angola the team collected millions of specimens—simply too much to deal with, and much of it still hasn't been. Then there was a project in Sulawesi organized by the Royal Entomological Society in 1985 to provide the first, and only, complete inventory of the insects of a tropical rainforest. Up to a hundred entomologists were involved. The canopies of trees were "fogged" with insecticide to bring down their hidden denizens. Rotting vegetation and soil were sampled—in fact every thing that could be sampled was sampled. Peter Hammond reckons he personally examined a million beetle specimens under his microscope. The idea was not to name everything, but to establish just how many species there were, to get some estimate of tropical biodiversity. The relevant specialist might be able to identify a genus, for example, and assign a dozen species to it, even if many of them were new to science. The exer-

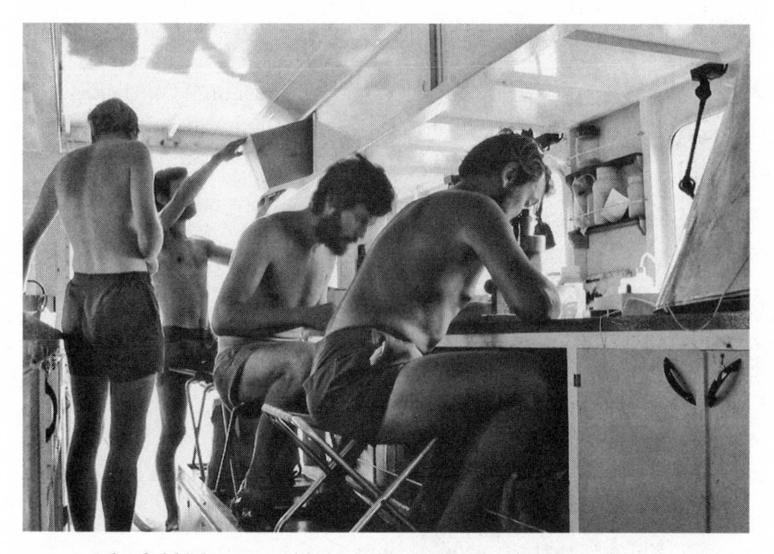

The field laboratory fabricated by staff on a shoestring budget that toured Africa collecting insects in the 1970s: systematic labour that is never finished

cise proved both that diversity was higher than anyone expected and that there was a host of unknown species living particularly in the lower canopy and in the litter layer.

Modern methods of characterizing species employ molecular sequencing to identify a characteristic part of the DNA of each species as a "bar code"; this speeds up the process of characterization and recognition. But this process leaves out everything else. Every species has its own tale, a story about how it earns its living, meets its mate or warns off its enemies: the interesting stuff. You don't understand London just by reading the names in the telephone directory.

As we have seen, beetles will eat anything—even slime moulds. In damp woodland you may notice pink or brown blister-like masses of spores sitting on rotten logs: these are the reproductive structures of some of the larger slime moulds. In the earlier phase of their life cycle they glide over the forest floor feeding on decaying vegetation: a slime mould in this stage of its life is like a patch of living snot. It looks like a

Agathidium vaderi, a small dark beetle named by Quentin Wheeler after Darth Vader, evil eminence of *Star Wars*

mixture of an amoeba and a fungus. Many people might regard this fascinating organism as one of nature's less appealing creations. In 2005 Quentin Wheeler and Kelly Miller named a series of slime mould–consuming beetles after President George W. Bush and some prominent members of his cabinet: *Agathidium bushi, A. rumsfeldi* and *A. cheneyi.* The press was tempted to draw a rather obvious conclusion from this. However, Quentin Wheeler assured me that he had been a Republican all his life, and that he was very fond of slime beetles. After all, in a work describing some sixty-five beetles he had named another species for his wife, and one after the Dark Lord of Sith, Darth Vader himself (*A. vaderi*). The authors were clearly fans of the *Star Wars* saga as well as the Bush government; the black outfit of *A. vaderi* was ostensibly reminiscent of the outfit of the Evil One. Not long after the publication of the new species Quentin was sitting in his office, carrying out his job as Keeper of Entomology, when the telephone rang. When he answered the call a voice responded: "This is the President of the

United States." He was about to respond with the usual "Oh yes? Well, this is Darth Vader . . ." when he realized that it actually *was* the President of the United States. He and his colleagues, Bush said, were honoured to be so immortalized in the names of beetles. *Multum in parvo* indeed.

7

Museum Rocks

There is nothing really special to look at in a Martian meteorite. A similar-looking piece of igneous rock might be picked up on the Isle of Skye or in Newfoundland. Yet these little fragments of rock have reached the surface of our world from Mars. They have ricocheted out into space, probably when the Martian surface was impacted by another meteorite, and then, after an interplanetary journey that might last for millions of years, they have landed on Earth. They must have survived passage through our own oxygen-rich atmosphere, where most meteorites burn up as "shooting stars"—thin bright flashes that flare and then are gone. In short, their survival is little short of a miracle. Martian meteorites are very rare—there are fewer than forty known specimens in total. They also change hands for thousands of dollars, so the lucky individual who discovers one of these rare stones might make a small fortune, literally out of the blue. They are also hard to recognize. What is required to find them is a place where the meteorites will lie undisturbed and unweathered over hundreds of years, and a pair of sharp and skilful eyes to pick out the precious article from a million other stones. These conditions are met in deserts of two kinds: the arid and the icy. A huge ice sheet receives meteoritic gifts from space, acting like a vast outstretched hand; the slow-moving glacier eventually gives them up again at its melting edge. If a scientist knows just where to go, he can discover natural concentrations of meteorites. Most of them will be the commoner kinds of metallic (nickel-iron) or stony meteorites, but in the Antarctic a few of the precious Martian examples have been recovered, the prize for financing a whole expedition. Martian meteorites have turned up particularly on the edges of the Sahara Desert.

I was driven over the top of the Atlas Mountains southwards to the town of Erfoud, in the Moroccan Anti-Atlas, where some of the latest finds have been collected. Ibrahim Tahiri runs a private museum on the edge of this small desert town, which comprises a few streets of undistinguished white buildings surrounding the traditional market, or souk. The new museum is a smart new warehouse full of fine cases and stands showing off a beautiful collection of fossils: giant trilobites and starfish enough to make a palaeontologist weak with acquisitive lust. Ibrahim Tahiri travels to the big rock and fossil show in Houston and knows exactly what a good specimen will make on the open market. He will greet you with a ready smile and proudly show off his treasures: he refuses to sell the cream of his collection displayed in the museum. Tahiri has also claimed new discoveries of Martian meteorites.

It is necessary to drive far out into the desert to see some of the places where meteorites have been found. Low hills display the geology completely; the strata follow one upon the other so that geological time is drawn out blatantly upon the hillside. In the carpet shops of Erfoud the piles of richly coloured rugs follow the same principles of simple stratigraphy. Between the hills, the low ground is often an almost flat plain composed of stony desert, not a tree or a bush in sight. Some areas are covered by migrating sand dunes, so creating a miniature version of the dune fields that cover huge tracts of the true Sahara Desert to the south. But the meteorites are collected on the arid wastes of stony desert: cheerless places that often seem to stretch onwards to infinity. I was curiously reminded of Thomas Hardy's description of Egdon Heath in *Return of the Native:* "like man, slighted and enduring, and withal singularly colossal and mysterious in its swarthy monotony." But, unlike anywhere in Wessex, by midday the *reg* shimmers in the heat, dissolving in the distance into a glassy mirage. The stony desert "pavement" comprises countless small rocks—and in some places they *all* look like meteorites. They are painted with desert varnish, a thin covering of iron and manganese that burnishes them purple-black. Some stones become pitted on the surface and then look superficially like nickel-iron meteorites—the class that includes some of the largest bodies to reach the Earth from space, but also many more commonplace examples. I am amazed that the local meteorite hunters can spot the difference between the stones and the meteorites—especially since some of the older meteorites have become secondarily varnished. But there are scouts out in the wasteland who scour the surface of the ground for precious exceptions to the common pebble. One of the most astonishing phenomena of the desert is the way that these Berber detectives suddenly materialize in the middle of nowhere. As the field party inspects an outcrop, a lanky figure dressed in a loose-fitting and frequently ragged robe will suddenly step out from behind the cover of a saltbush. He will lope towards the party at his leisure, and then produce an ancient cloth bag containing treasures, smiling enthusiastically all the while and exposing one or two surviving teeth. Although there will be trilobites and other fossils, pride of place will go to the meteorites he has managed to find among the millions of undistinguished stones. Some of them will be genuine. If your interpreter asks the Berber how long he has been out in the desert, the reply will come back "Many days," accompanied by a vague wave in the direction of distant hills. But anything of real worth will find its way back to Tahiri. Even in the remoter parts of Morocco the middleman reigns supreme.

I have known a succession of meteorite experts at the Natural History Museum. The senior member, Bob Hutchison—familiar to many as "Hutch"—was a short, bearded Scotsman with an infectious, barking laugh, and an admirable set of principles concerning equal opportunities and the rights of Mankind. His successor, Monica Grady, is a large and sociable (rather than socialist) woman, as prone to wild laughter as Hutch, although equally high in principle, but perhaps from an attachment to the Church rather than to Keir Hardie. She radiates bonhomie like a perfume. From its early days the Museum in London was central to meteorite research. Lazarus Fletcher, the Keeper of Mineralogy in 1881, published a pamphlet as a catalogue of the collections; in 1923 George Prior, Fletcher's successor, published the first of the global catalogues, and the *Catalogue of Meteorites* has been continually revised ever since. Monica Grady produced the fifth edition in 2000. It lists 22,507 authenticated and catalogued meteorites, which proves just how rare those Martian examples are. All meteorite researchers refer to the *Catalogue* for their taxonomy—the numbering system that ensures that each meteorite is uniquely identified. The Martian meteorites are known as SNC meteorites, which is shorthand for Shergottites, Nakhlites and Chassignites—the three known varieties of these alien rocks—and one example where an acronym is probably preferable to the original. The ungainly names are derived from the localities where the first typical examples of the meteorite were found: so Chassignite is named after Chassigny, in Haute-Marne, France, where its rare type was discovered in 1815. The Martian signature of these meteorites was suggested more than twenty years ago when it was discovered that the noble gases were present in them in proportions very like those determined by the Viking spacecraft directly from the Martian atmosphere. Since noble gas elements, such as Argon, Xenon and Radon, do not react with other elements, they are like an unshiftable family heirloom, and point the finger firmly at the ancestral home. Subsequent work on isotope ratios of several other elements has also identified a good Martian signature, and few scientists would challenge the origin of these meteorites today.

Monica Grady became scientifically engaged with claims that there were traces of life in the rock samples from Mars. The claims were twofold: that there were fossils in the meteorite samples of tiny rod-shaped objects that might be "bacterial" (so-called nanobacteria, because they were smaller than usual bacterial species), and that the chemical signatures of elements or compounds found in the meteorites were likely to be associated with organic activity. Much brouhaha has been generated by claims that the presence of life on Mars has been fully proven; an endorsement of the theory was even given by President Clinton. A cynic might have said that these spectacular reports might have had something to do with the quest for continued funding by organizations like NASA. A few meteorites have figured prominently in the arguments—like the Shergottite, Catalogue number ALH 84001. Even though Earthling bacteria are a few thousandths of a millimetre long,

Left: The Martian meteorite ALH 84001 (which has had some material removed for analysis)
Below: Micrograph of the supposed "fossil" from ALH 84001—not widely accepted as such

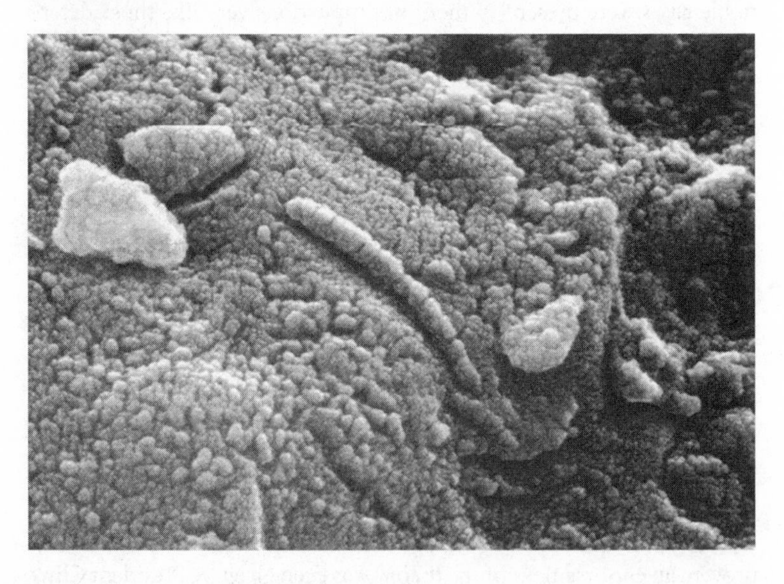

the minute rod-like structures claimed as nanobacteria from ALH 84001 were a whole lot smaller. The renowned expert on early fossil bacteria Bill Schopf, from the University of California, Los Angeles, declared that he did not believe that the tiny traces were relics of life forms. Then there were discoveries that organic molecules usually associated with biological synthesis could be found in a few SNC meteorites, and their concentrations were greatest *inside* the meteorite, ruling out contamination. There is nothing unusual about carbon in meteorites—indeed, well-known varieties are called carbonaceous chondrites for reasons that do not require spelling out. It has long been known that these meteorites include traces of amino acids—one of life's "building blocks." But these varieties of meteorites are also thought to be very primitive—the kind of basic matter from which the solar system itself was formed. It remains a matter of controversy whether crucial organic molecules for creating life were delivered to Earth by extraterrestrial messengers, or whether the chemistry happened here on the home planet. One of the most exciting things about studying meteorites is that they provide a kind of telescope to see back into the earliest days of the evolution of Earth and its neighbours. Perhaps the wisest option is to keep an open mind about life on Mars, now or in the past, but be very sceptical about those alleged tiny fossils. With more probes beaming back information from the red planet itself, the question should be settled once and for all within the next decade or so.

Improbable though it may seem, there is a crossover between meteorite studies and my own favourite animals: trilobites. In southern Sweden numerous vast quarries have been opened up to exploit limestones of Ordovician age. These grey rocks have been used extensively as flooring in churches, or, polished, as an ornamental "marble." They can be distinguished from a hundred and one other ornamental limestones because they carry numerous cross-sections through fossil shells of nautiloid molluscs, which are long cones shaped like the wafers that hold scoops of ice cream; these fossils are divided internally into chambers, so they are very distinctive. The same limestones frequently yield trilobites, which have been used to date the rocks; I know some of them as well as I do my old friends. Because of the great geological stability of eastern Scandinavia, nothing much has happened to these limestones in the 470 million years since they were laid down under a shallow sea; they still carry their fossil evidence in perfect condition. But a trilobite innocently scurrying over the muddy sea floor back in the Ordovician would have had its peace shattered by the arrival of showers of meteorites. Bob Hutchison and his co-workers, Drs. Schmitz and Bridges, have managed to identify "fossil" meteorites in these Ordovician deposits. Because of the way the limestone is quarried, bed by bed, an investigator can crawl over wide, flat areas in some of the quarries, which are effectively exhumed ancient sea floors. Although not exactly common, old meteorites are probably easier to find in these quarries than their equivalents in the Sahara Desert. They show up as dark, walnut-sized blobs on the pale surface of the limestone, and are often surrounded by a rusty halo. Not much of their original chemistry is preserved, but the unusual mineral chromite has survived through hundreds of millions of years to retain an unmistakable signature of the common class of meteorite known as L chondrites. It seems that during the Middle Ordovician the world passed through a massive meteorite "cloud." Schmitz, Bridges and Hutchison have found evidence of more than twelve meteorite showers at that time. It may well be that this will prove to have been a worldwide phenomenon, although few places are as ideally suited to finding "fossil meteorites" as southern Sweden. Maybe this global meteorite shower will in turn be connected with a dramatic changeover in the trilobites and other animals that happened about the same time, when many new forms appeared in the fossil record . . . The interesting thing about scientific questions is that one bounces off another like a series of cannoning billiard balls—and just occasionally one finishes up in the pocket.

Mineralogists appear to be rather a sensible lot compared with biologists and palaeontologists. In my attempts to extract oral history from my former colleagues, it proved remarkably easy to get revelations from those who worked upon organisms—whether living or fossil. Quite a few of these anecdotes involved sexual relations between people who I had no idea had had any relations at all, let alone carnal ones. I was already aware that Dry Storeroom No. 1 was a secret trysting place. That dry old sunfish had witnessed many a lubricious episode. I later heard that the attic floor of the old Entomology Block had mattresses deployed to help the entomologists with their studies of the human genome. It seems that the Dark Room was often locked from the inside because of unforeseen developments. Until I interviewed the protagonists, I had no idea of the lustful tendencies of experts on weevils, toads or brachiopods. By contrast, the mineralogists seem to have been altogether better behaved. This may have something to do with the fact that Mineralogy is the smallest department, and the most overseen. Until quite recently, all the post addressed to the scientists was opened in the Keeper's office, which was certainly inconvenient for anyone trying to arrange a secret assignation. Mineralogists also tend to be the more mainstream scientists. They are the ones that wear the white coats, and hide away in the basement while reading dials from sophisticated machines. Only a few of them have gone mad, and many of them have lived blameless lives in the single-minded pursuit of mineral excellence.

Minerals do have systematics and taxonomy, just like organisms. Like animals and plants, they are classified into species, although there are far fewer species than in the biological world. The old mineral gallery in the Museum still has the crystal species laid out in an order dictated by their chemistry. The very word "mineral" tends to conjure up a picture of some beautiful and exotic crystal in a display case, but the bulk of minerals are ordinary components of common rocks. A lump of granite is a mass of minerals locked together in a three-dimensional jigsaw. This is an easy fact to verify while one is waiting in a queue in the bank, where polished slabs of granite seem to be invariably displayed on columns and counters. Pale pink or white feldspars speckled among quartz give the rock its dappled texture: there will be three or four mineral species on display. A mineral species has two diagnostic properties: its chemistry and its crystallography. Even if two minerals have the same chemical formula, they can have more than one name if they display more than one fundamentally different crystal form in nature; the simplest example is diamond and graphite, which are both forms of the element carbon, but could scarcely look more different. Because there is a limited list of chemical elements and they can combine with one another only in specified ways, the number of natural mineral species is far from infinite, although there is no sign that science is running out of new discoveries just yet. Many elements are rare in nature, so minerals containing them will also be rare. The appropriately named rare earth elements (with strange names like Yttrium) are very seldom found either singly or in quantity, although they have become very important in geochemistry and can almost be counted atom by atom with modern instruments. Despite their rarity, rare earth elements are useful. Yttrium is used in the high-intensity lamps of cinema projectors, for example. Some elements, like the noble gases we have already met, are so snooty that they won't combine with anything else except under very exceptional circumstances. By contrast, a few elements—silica, aluminium, oxygen, iron, calcium, carbon, hydrogen—are so abundant that minerals combining some of them in various permutations are found practically everywhere. Silicates—compounds of common quartz—are especially fecund and various because silica molecules can join together in all manner of different ways to form sheets or nets that welcome in the other common elements. Many families of minerals, with names like pyroxenes, feldspars or amphiboles, are silicates that share a common structure. When pure and well formed, a given mineral will usually have a characteristic crystal shape, and often one we find beautiful. This crystalline perfection is a reflection in the hand specimen of the way atoms are stacked and arranged right down at the atomic level. The mineral mica, for example, breaks up into thin sheets if plucked with a fingernail, and the silicate molecules of which it is composed are also arranged in sheets. Common rock salt crystallizes into cubes, and the elements sodium and chlorine of which it is composed are also arranged cubically. The macrocosm mirrors the microcosm.

Mineral structure at the molecular level was first investigated by the great X-ray crystallographers: Sir Lawrence Bragg and his successors. The X-rays sneak in between the lattice of atoms to produce characteristic arrays of diffraction patterns related to the way the atoms are stacked. These modern investigators built upon centuries of work by early mineralogists. One could argue that science itself grew up among the alembics of the alchemists; on the bench in front of these arcane wiseacres would have been elemental sulphur crystals, or minerals with ancient names like realgar or orpiment. Some of these minerals were derived from smoking fumaroles around active volcanoes such as Mount Vesuvius, the very distillations of the bowels of the Earth. Between the seventeenth and the twentieth centuries the chemical elements were teased out of their compounds one by one. There was a metaphorical dimension to the discovery of a deeper truth about matter, which was mirrored in the geological depths from which many of the minerals originated. Mineralogy came from this ancient tradition, and the modern science gradually shed the esoteric baggage of its forebears. One of the earliest no-nonsense science* books was Georgius Agricola's *De rerum metallica* (1555), a practical guide to mineralogy and the arts of mining. It remained useful for several centuries. As knowledge of chemistry and the elements developed, the old furnaces of the alchemists were replaced by the blowpipes of the assayers, and then by the batteries of reagents—strong acids, solvents and poisonous cyanides—used by "wet" chemists, the men and women who use test tubes and titration to identify the composition of a mineral. Even today there is still a "wet lab" in the Natural History Museum used to identify certain of the lighter elements. But most of the routine work of assessing chemical composition of the majority of minerals is now entrusted to high-tech equipment: electron probe microanalysis and ion probes can work on tiny quantities of material, even a sample only five microns across, that is, five-thousandths of a millimetre, plucking out and sorting its atoms to an accuracy of picagrams (that is, a million millionth of a gram). There is something almost mystical about these kinds of figures, something that should inspire in the ordinary person a feeling not unlike the awe felt by an initiate wandering into the alchemist's lair. But as the figures are derived from machines, faced with dials and plasma screens that are familiar from a hundred films featuring the scientist at work, somehow the achievement of such accuracy can be taken on the nod. It is remarkable how the remarkable has become unremarked.

New minerals are still being discovered regularly, and part of the job of the Mineralogy Department is to describe them chemically and crystallographically—and only then to provide a new name. Names have to be approved, and there is a special commission of distinguished mineralogists to make sure that something claimed as new really *is* new. The International Mineralogical Association has a Commission on New Minerals, Nomenclature and Classification to vet the validity of

*I am aware that using the word "science" here might go against the cautions of modern historians to avoid anachronisms. It would have been even more anachronistic to have used the word "geology" as that scientific concept did not appear until the end of the eighteenth century, as Martin Rudwick has explained in detail in *Bursting the Limits of Time* (2006).

new discoveries. A couple of dozen species might be approved in any normal month—nothing to impress a beetle enthusiast, of course, but still proving that there is much to discover within the Earth. Chris Stanley tells me that most of the new minerals he has described and named are not very exciting to look at, often no more than a dusting of tiny crystals. It is no wonder they were undetected by earlier mineralogists. He and his late colleague Alan Criddle have named nearly a hundred new mineral species over their careers. It is only thanks to the delicacy of the new technology that they can be characterized so accurately.

Occasionally, Chris gets a surprise. He showed me part of a borehole core from Serbia, something substantial enough to toss from hand to hand. The borehole had been put down through a thickness of volcanic rocks. This particular piece of core consisted largely of a milky-coloured material which proved to be a completely new mineral. It contained a high proportion of the lightest metallic element of them all, lithium, which makes it a most surprising find. In 2007 it attracted press attention because its chemical formula matched that of Green Kryptonite, the only substance to which Superman was vulnerable. It will be called Jadarite, after the place of its discovery. It must be published with details of its chemical formula, crystal structure and atomic proportions before it can be considered valid. As for naming, this geographic formality is common among new minerals; a site where a mineral species has been discovered will have "-ite" tacked on to the end of it to give the mineral name—not very imaginative, perhaps, but easy to follow. Quite frequently, a mineral will be named after a distinguished scientist, as in Zinnwaldite (after Dr. Zinnwald),* so this is not unlike celebrating a botanist in a plant species name. Several hundred new species of minerals have been named in the Mineralogy Department over the last decade. More confusing is the fact that rock types are very often -ites as well, ranging from andesites to tholeiites, and sometimes these are named after localities, too (like Chassignite above). Such rock types are mostly collections of minerals en masse. Their definitions are much laxer than those of minerals, so there are several varieties of

*To give another example, in 2002 Alan Criddle named a mineral Frankhawthorneite, you may guess after whom.

The hand specimen of the mineral Jadarite, recently named: chemically it is sodium lithium boron silicate hydroxide; it achieved unusual prominence in the media because of its resemblance to "green kryptonite"—the only substance known to weaken Superman.

andesite, but they are nonetheless part of the common language of geologists, just as the names of species help communication between biologists. Nomenclature is important.

Rock is the real stuff. This is proved by the voyage to the Moon: such an expedition must be validated by the collection of rock samples. All that trouble to acquire something that looks so ordinary—but you cannot argue with it, because it is solid as a rock. I recall the excitement when Moon rock appeared in the Museum in the 1970s—here was evidence you could really believe. We all looked at the small tube containing the sample with respect. It is no coincidence that Jesus' most reliable disciple was called Peter, "the rock" on which the Church was built. The Natural History Museum keeps historical collections of rocks. Early expeditions risked nearly as much to bring back these unglamorous lumps as did the first explorers on the Moon; they, too, collected hard and most incontrovertible testimony to their boldness. Part of the Mineralogy Department is located in the basement and has been labelled "Miner Alley." One side of the corridor has high glass cases which include splendid crystal specimens and old optical instruments, all burnished brass and elaborate screws used to raise and lower their universal stages. On the other side there is a long rank of old-fashioned cabinets containing rocks—historically significant rocks numbered in sequence and purchased with the privations or even death of forgotten field staff. Here is the sad booty of Captain Scott's Antarctic expedition. The oldest collection is probably that made by Sir William Hamilton, envoy to Naples and cuckolded husband of Nelson's Lady Hamilton—and also a pioneer archaeologist and writer of the magnificent *Campi Phlegraei* celebrating the wonders of the Bay of Naples. Here, too, is preserved the geological collection of the Matthew Flinders expedition to Australia in 1801, probably the first rocks ever brought back from that continent. These rock types can easily be re-collected again today, but then they were the first blobs on a geological map. Now I guess that these drawers are mostly opened by historians rather than by geologists. No doubt if Moon travel ever became routine the first Moon rocks would become historical curios in their turn. The Vesuvian rocks of the Monticelli collection are the exception: they are still consulted, because of the precision of the times, dates and places of their acquisition by their careful collector. Geologists who want to know how magma evolves during a volcanic eruption have here a unique database of past crises.

Even the way that rocks are studied—the science of petrology—has changed repeatedly. When I was a student, the chief tool for studying rocks was the prepared thin section (like the one on colour plate 13). The rock was trimmed of a thin slice, which was mounted on a slide, and then ground still thinner until its component minerals could be studied under the microscope by shining a directed beam through them. The mosaic of minerals so revealed had a chequered beauty, like a brilliantly coloured abstract painting, especially when viewed under conditions of polarized light. We were taught to identify minerals by their optical properties, as our teachers had themselves been taught by the Cambridge legends of the light microscope, Professors Tilley and Harker. There was a certain satisfaction in learning these mineralogical skills, and some intellectual satisfaction to be had in linking our identifications with the chemistry of the rocks themselves. Ph.D. students cut literally hundreds of thin sections to get their data on composition of magmas, or the temperature and pressure conditions to which a particular gneiss had been subjected when it was deep within the Earth's crust. A modest number of thin sections continue to be sliced today for petrological microscope study. Opaque minerals, such as those in meteorite "irons," are still studied from highly polished surfaces. But the sophistication and convenience of the analytical machines have transformed the study of rocks, so that thin sections now play a lesser part in most research. Old timers will grumble, as old timers will, that the youngsters "wouldn't know an adamellite if it hit them in the face." But scientific instruments change, just as research priorities change—though I cannot see an old electron probe making as attractive an exhibit as those beautiful essays in tooled brass and hand-ground lenses that line the walls of "Miner Alley."

Alex Ball and Terry Williams are in charge of the Kingdom of the Machines that now occupies the basement area under the Earth Science galleries, and they will take you from room to room with proprietorial pride. The first thing you notice is how clean everything is compared with the average cluttered Museum office. Record books are neatly filed away above wiped-down benches—the word "shipshape" comes to mind. The Natural History Museum is rather well off in the latest technology, although, as Alex remarks, you have to run hard just to stay up with the leaders. Many university departments have to make do with one analytical machine that has to be constantly reprogrammed, but the Museum can find a machine to suit the job in hand, which means going into one dedicated room or another. Before the basement was commissioned, the machines were dotted about the Museum in obscure places. The first scanning electron microscope (SEM) was up and running in the Museum (1965) even before I joined the staff, thanks to the efforts of Ron Hedley, who later became the Director. He recognized its importance as a tool to see fine structures more clearly than ever before,

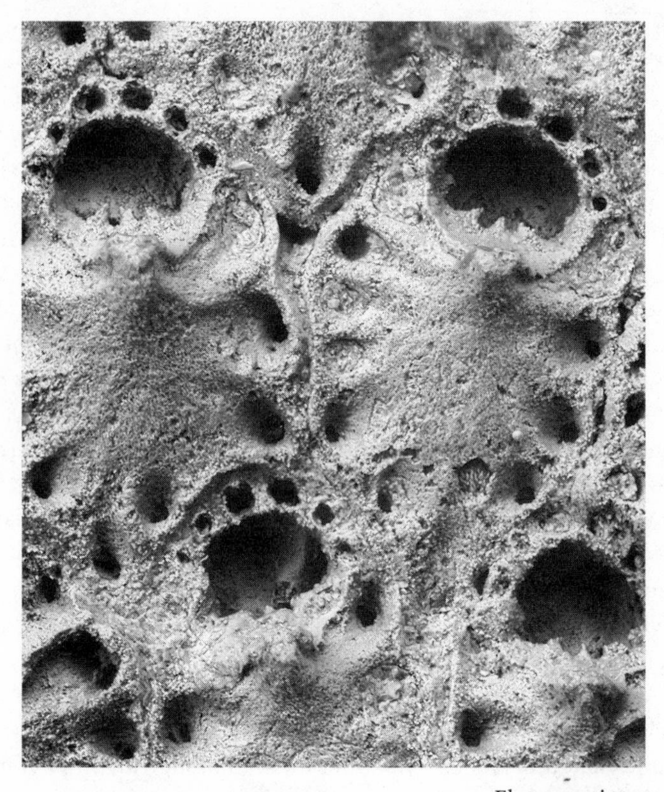

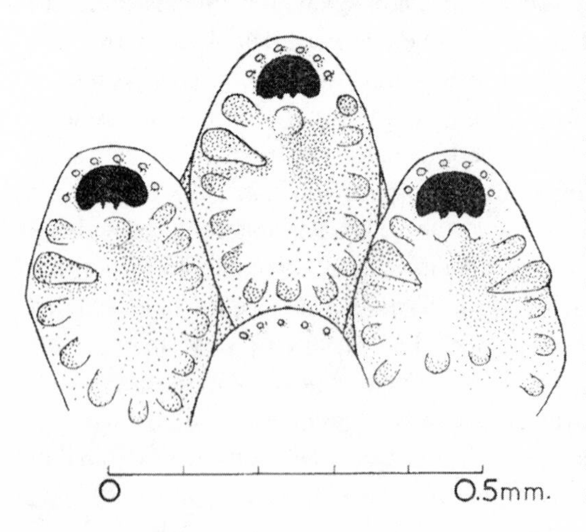

Electron microscopy is located in the Mineralogy Department. Images provide unrivalled details, even of fossils. Here (*above*) a photograph of the Cenozoic bryozoan (*Exochella jellyae* Brown, 1952) can be compared with the drawing of the holotype of Brown made in 1952 (*left*).

by using electron beams that could discriminate detail much more finely than the traditional light source. It is easy to forget the astonishment of being able to see for the first time the eye of a fly, or those "hairs on legs," with such precision.

This tool transformed the study of some animal and plant groups. Paul Taylor, the doyen of the bryozoans, is always stressing the beauty of his tiny, water-dwelling and mat-forming colonial animals—and he is right. Before the SEM these animals were usually illustrated by drawings, which varied greatly in quality. Now the exquisite and delicate patterns made by the colonies and the ornament of the little boxes in which each individual of the colony lived are both scientifically accurate and a delight to the eye thanks to the electron microscope. The scanning microscope was soon complemented by the transmission electron microscope, an instrument that has transformed our understanding of the organelles inside living cells and the way those cells collaborate to make tissues. I used the scanning electron microscope quite early in my Museum career. I always felt like a *real* scientist when I sauntered off to the machine. I soon had images of trilobite larvae a millimetre long blown up to the size of a small lobster, displayed on the green plasma screen attached to the microscope. The process of photography was slow because specimens had to be coated in a fine layer of gold to allow the electrons to "take"; later machines allowed photography of uncoated specimens, and these are still in use today.

So machine was added to machine, one by one, each with its own acronym, each doing a somewhat different job. For example, many mineralogists were not interested in getting good pictures so much as in analysing elemental composition. Much of the serious research money in the Museum came to be spent on this hardware. The latest version of the electron microprobe cost half a million pounds, but it saves much staff time because it is automated to assay for four elements simultaneously, and gradually works through an elemental "shopping list." Mineral samples are analysed from polished surfaces of rock slices, or they are sometimes powdered, or put into solution, depending on the technique involved. Machines have to be continually updated as new levels of accuracy of measurement are achieved—so, for example, they can now routinely focus on minute areas just a thousandth of a millimetre across if required. The mind soon reels when confronted by the variety of "kit" in the Kingdom of the Machines, but I was happy to see my old machine still sitting in one of the rooms.*

Some mineralogical investigation is closer to industry than most of the research that goes on in the Natural History Museum. There is money to be made from knowing about how ores or gemstones form in nature. Many valuable metal ores are associated with special rocks known as volcanogenic massive sulphide (VMS) deposits—and, as with SNC meteorites, it might be better to stick with the acronym for reasons of brevity. These are interesting and unusual rocks because most of them originated in ocean basins where cold seawater meets hot fluids from deep within the Earth—yielding sulphide deposits rich in zinc and copper. The metals are dissolved in water as hot as 380 degrees centigrade, and when the metalliferous liquor hits cool sea temperatures the solubility of the valuable ions falls dramatically. The resulting massive metal sulphides can be thick enough to support large quarries. The sites where these strange deposits form today typically lie along the mid-ocean ridges, which is where the lithospheric plates of which the world is made are slowly, slowly moving apart. Heat rises from the interior of the Earth along the ridges, bringing treasure—not just zinc and

*For those with a technical turn of mind, here is a selection of the devices that lurk in their separate rooms: energy dispersive X-ray analysis (EDXRA) identifies elements from their X-ray spectra; the field emission SEM is able to discriminate two points only two nanometres apart, and can be used to look at dust or the finest of mineral fabrics; the confocal microscope uses laser light to produce images with unprecedented depth of field; the Fourier transform infrared microscope (FTIM) produces infrared spectra characteristic of different materials—so, for example, fake amber can be quickly distinguished from the real thing, because the latter has a very distinctive spectral profile. Across the corridor the chemical analysis division boasts yet more machines in rooms separated from those equipped with the more familiar battalion of reagents and glassware that recall the archetypal "chem lab" of my schooldays. Quite a lot of these laboratories' efforts are directed towards getting minerals into solution so that they can be analysed by the machines in the adjacent rooms. These include "kit" such as the inductively coupled plasma mass spectrometers (ICPMS), which are used to measure trace elements and isotopes with an accuracy that boggles almost any part of the anatomy that is bogglable—if the blurb accompanying the machine is to be believed, this means a few parts per quadrillion, or 10^{-15}. Measuring such minute quantities of rare elements is important in identifying the sources of rocks; for example, igneous rocks derived from melting of the mantle will have a distinctive elemental signature different from

copper, but silver and gold as well. Sulphurous fluids belch out of vents known as "black smokers" which track the ridges on the ocean floor—and build up dark chimneys of iron pyrites out of sight of the sunlit world above. In the same hidden world thrives a whole ecosystem of bizarre animals, whose economy is based upon the sulphurous exhalations of the smokers rather than upon sunlight. There are shrimps that cultivate sulphur bacteria, or scrape them from the walls of the chimneys. There are vestimentiferan worms that house bacteria in their guts and form thickets of tubes. There are giant clams. It is an extraordinary world, and one that can be visited only in special diving craft like *Alvin*, the U.S. Navy's deep submergence "submarine" that can withstand the enormous pressures at depth. As far as mining is concerned, it might as well be on the Moon. Many of the "fossil" VMS deposits are associated with former volcanic island arcs, like those around the Pacific Rim today. They are preserved only in accessible locations on the continents because of the inexorable movement of the tectonic plates around the Earth. Pieces of ancient ocean floor finish up incorporated into the continents, and then they are available to miners. Geologists recognize an appropriate tectonic setting, typically where ancient sea floors have been subducted away, and this makes for excellent prospecting.

VMS deposits are important sources of industrial and precious metals. Gold may be the legendary lure for the adventurer, but more

that of rocks derived from melting of the crust—even though they might look very similar in hand specimen. We have seen previously how signatures can be recognized for Martian rocks. It is paradoxical that to understand some of the biggest questions about the origin of planets or the interior of the Earth samples have to be studied at the smallest scale that technology allows. Although one of the ICPMS machines requires the material to be in solution, the laser ablation machine uses a precisely focussed laser beam to sample a small area of a solid sample. Some important elements are not satisfactorily assayed by these machines—especially carbon, hydrogen, nitrogen and sulphur—so for these there is another special apparatus (not surprisingly called the CHNS analyser)—which essentially works by burning the sample and measuring the product of combustion. Any element that needs to be analysed and measured can be, down to the atomic level. The black boxes produce a kind of chemical black magic. It is odd to reflect that this humming world of plasma screens and analysers is the logical successor to the old assayer with his blowpipes and porcelain plate to look at the "streak" of a mineral. It is sometimes hard to remember that these machines are our servants and not our masters.

commonplace metals like copper may prove to be more relevant to the future of the world. It is a curious thought that the electrical wiring in our homes might have started out on an ocean floor millions of years ago. Richard Herrington is the Museum mineralogist and metallurgist with a special interest in VMS formation. He is full of enthusiasm for heavy dark rocks made mostly of massive iron pyrites—iron sulphide, otherwise known as "fool's gold"—not one of nature's most elegant productions to my eye. But his work with Crispin Little of Leeds University and Russian colleagues has cast a brilliant light on ancient oceans, showing that the community of life around "black smokers" has been established on Earth for hundreds of millions of years. These scientists have been working in the copper/zinc mining districts in the Urals, especially around that city whose name leaves little to the imagination, Magnitogorsk. Its famous ice hockey team is the best in Russia and leaves even less to the imagination—they are called Metallurg Magnitogorsk. A glance at a map of the world will show how the Ural mountain chain snakes across the centre of Russia from Novaya Zemlya in the north, by way of some of the new republics, all the way down to Kazakhstan in the south. It is a huge wrinkle on the surface of the globe. This shape alone would suggest to a modern geologist that the Urals represents the aftermath of a vanished ocean—for linear mountain chains like the present Himalaya are thrown up in slow convulsions of the Earth's crust when continents collide. Just the place for VMSs. The eastern and western halves of Russia came together several hundred million years ago at the end of the Palaeozoic Era. Prior to that they lay on separate plates, and an ocean with offshore islands lay between them—one that was consumed by a subduction zone on its western side. This eventual welding of the two continents together was a long-drawn-out process, and several island arcs found themselves plastered on to the nascent Uralic chain many millions of years before the continents themselves collided—they were the advance skirmishes before the main onslaught.

The earliest vent fauna yet known to science has been recovered from one of these precocious slabs of ancient ocean floor near Yaman Kasy. It is Silurian in age, about 430 million years old. The fossil tubes of vestimentiferan worms are about the dimensions of long macaroni—

but transformed into a heavy, blackish metallic-looking mass by the iron pyrites engulfing them. To a seasoned palaeontologist Yaman Kasy provides the most improbable location ever for finding evidence of past life. Normally, volcanic rocks are completely devoid of any organic remains, except in very rare cases where animals have been completely overwhelmed by an ash fall, and the idea of finding fossils in solid lumps of iron pyrites would have most of us slapping our thighs and guffawing. But then, vent faunas are a unique kind of biology, and the rulebook has to be thrown away. Little and Herrington have also found large masses of fossil vent fauna in VMSs of Devonian age in the large copper workings near Magnitogorsk; these are about fifty million years younger than the Silurian occurrence. This fauna includes a large clam and a snail as well as the ranks of tubes belonging to the specialized worms. It is a disappointment to me that no trilobites have been discovered yet in any locality—I have fantasies about them occupying the specialized bacterium-grazing niche that the crustaceans occupy today.

It is an extraordinary tribute to the opportunism of living organisms that even this strangest of habitats should have been colonized so early. There remain some controversial aspects to the interpretation of the fauna: some workers have claimed that the animals were not living as deep as they are today. And Richard Herrington and his colleagues are still putting together the story of the appearance of the Urals, which certainly involves several separate phases of island arcs being accreted to the edge of the ancient continents (arc "docking"), and may also involve major fault movements that slid chunks of ground lengthways along the mountain chain. Differences in metallic composition from one ore body to another can be explained by different origins within the vanished ocean. Richard is working on a complex story of alteration products of VMS by the actions of hot fluids that modify the chemical composition and mineralogy of ores after they are formed. It is a convoluted story, but at the end awaits a better understanding of how our Earth is put together, as well as new sources of wealth. Minerals are the end product of natural cookery in the cauldron of the Earth—and the recipe can be reshuffled several times. Who knows if some tiny crystals may yet prove to be a new species, having waited two hundred million years to receive the blessing of a name?

There are even odder rocks under study in the Mineralogy Department. Carbonate rocks like limestones and dolomites are abundant sedimentary rocks. They form the Cretaceous white cliffs of Dover in England, and the Permian Capitan reefs of the Carlsbad Caverns National Park in the United States. Limestones are just about everywhere, not least as building stones—they flavour much of France, and impart that soft, golden glow to Bath. Many of these rocks were originally formed from chemical precipitation of carbonates from seawater, often associated with bacterial activity, or from the fossils of animals with shells made of calcite or aragonite (these are common forms of calcium carbonate). They are where a palaeontologist likes to go to work; they belong to the world of sea, life and sunlight. Imagine, then, an eruption of carbonates from a volcano. Yet this is exactly what happens when Oldoinyo Lengai, an active volcano south of Lake Natron in the eastern Rift Valley of Africa, bursts into life. The geologist Arthur Holmes described the 1917 event thus: "The volcano suddenly burst into an eruption which lasted several months and shrouded the country for miles with soda-permeated ash. With the first rains the water holes became fouled with bitter salts and many herds of cattle died through drinking from the contaminated pools. Lava that flowed down the slopes of the volcano cracked into irregular blocks looking like grey cement." It is a weird, bleak place, where all the usual rules of rock behaviour are suspended. These volcanic rocks are known as carbonatites, and they have been studied at the Natural History Museum by several generations of scientists. Some of the earliest samples were collected by the Overseas Geological Survey early in the twentieth century and sent back to England, where Campbell Smith, who eventually became Keeper of Mineralogy, recognized that they were erupted as lavas rather than being limestones caught up in an eruption. Then an incorrigibly amiable scientist, Alan Woolley, subsequently made carbonatites his life's work, and continues to work on them in retirement, while Frances Wall carries forward their study into the next generation.

Geology evidently requires patience. Nowadays carbonatites are recognized all over the world, from different geological ages, some of them even in very ancient rocks dating from the Precambrian. But as more has been learned about them, the more interesting they prove to be. Frances Wall glows with magmatic enthusiasm on the subject, while Alan Woolley erupts in chortling flows of words. Some of the most curious carbonatites have proved to have diamonds in them. This tells us that they originate not from the higher levels of the crust where sedimentary limestones are normally found, but from very deep down inside the Earth—even from the mantle, where temperatures and pressures are enough to convert the dross of carbon into sparklers, albeit rather small ones. It seems that this kind of carbonatite is extruded from "tubes" punched deep through the lithosphere. Occasionally, these conduits carry up lumps of dark mantle material, displaced messengers from the depths, as happens near the Monticchio Lakes in Italy. Quite commonly, little balls, or spherulites, of graphite about a millimetre across can be separated from the samples, and these are believed to have originated from the parent magma at a temperature about 700 degrees centigrade. Frances has been pursuing carbonatites in fieldwork in the southern Naratau Mountains of Uzbekistan; this undulating area of arid nothingness at the western end of the Tien-Shan would tempt only a geologist or an ascetic seeking mortification of the flesh; or possibly someone in search of the way diamonds form in nature. Once samples had been collected, experimental work could continue back at the laboratory: if a sample of the Uzbek carbonatite rock is seeded with graphite and then the mixture is melted under great pressure, the conditions needed for diamond formation can be replicated. It turns out that diamonds precipitate out at a pressure of 7Gpa and a temperature range of 1,200–1,570 degrees centigrade. It seems that if conditions were extreme enough diamonds might be formed within the carbonatite magma itself deep within the Earth.

Yet another bleak area yielding carbonatites embraces the boundary between Finland and Russia: the Kola Peninsula. Endless forests of dwarf conifers and glacial lakes give way to boggy permafrost, the kind of landscape that swaps freezing winter wastes for a brief episode dominated by summer mosquitoes. In this remote region carbonatites were formed in a different way from "pipes" penetrating deeply into the Earth's innards. Instead, they formed at depth in the Earth's crust during a phase of intrusion of a great body of hot igneous magma, one that was already particularly enriched in sodium and potassium. A kind of distillation of magma operates here. As more and more minerals are crystallized out, the remaining magma becomes progressively richer in the "leftovers": volatile components, or elements which have particular affinity for sodium, or even awkward elements that are reluctant to fit in anywhere else. It is a process of progressive refinement, and—like the business that goes on in an oil refinery—several different products can be produced according to local circumstances. Many volatile fluids invade the country rocks, where they crystallize slowly in fissures, to form *pegmatites*—the source of some of the most perfect and beautiful minerals. Bathed in fluids and gases, these crystals can slowly grow to perfection, as if to please a godly gemmologist, every face perfectly sculpted, and their design naturally determined by their constituent atoms. But because these are carbonatite magmas, the pegmatites contain minerals that are found nowhere else on Earth—they have been enriched in rare elements that provide the ingredients for a host of unusual chemical compositions. The elements concerned include names such as Lanthanum and Yttrium—the rare elements mentioned above. The bleak wastes of the Kola Peninsula provide an unrivalled diversity of strange species of minerals, a gem-hound's dream. Some of them are as beautiful as rubies, though far, far more rare. Frances Wall and her colleagues have produced a catalogue of these species, a concatenation of odd names like Scherbakovite or Kovdorskite (colour plate 15), which immortalize a bunch of Russian mineralogists and place-names. When the Natural History Museum started working in the Kola, the local Russian mineralogists compiled a shopping catalogue of minerals available there: not one of these minerals was represented in the collections in London. Now we have at least a selection; everyone benefits from international collaboration. Nor are minerals containing rare earth elements of purely academic interest. These elements are of increasing importance in the electronics industry, and wise money is moving into this area, with the unfailing instinct that money always has for the coming thing.

The public gallery displaying the minerals is unusual in the Natural History Museum in also housing the bulk of the collections. For the rest of the Museum departments the collections are so vast that only a tiny part of them is actually in the galleries. Furthermore, every gallery

The mineral gallery of the Natural History Museum in about 1920. The systematic arrangement of minerals is preserved today, even though most of the exhibits elsewhere in the Museum have been transformed from their early-twentieth-century state.

elsewhere in the Museum has been revamped over the last couple of decades, but the mineral display at the eastern end of the first floor has been preserved in something like its original state. The Victorian Society is delighted. It is an airy space, well lit from the generous windows, and with glass-topped cabinets running in ranks transversely across the gallery, each of which includes a fine selection of specimens. The arrangement of minerals in the cases is by natural "families" of minerals—so the sulphides will be found together, as will the native elements like gold and copper, or the oxides, and so on. It is a teaching collection in a way that no longer exists elsewhere in the building. An eager visitor might spend weeks in here learning, and would emerge at the other end as something of a mineralogist. Underneath the glass cases are ranks of polished drawers locked safely away that contain the systematic collections. Under silica, for example, there is every variety of quartz from brilliantly transparent rock crystal, through yellow citrine or delicately pink rose quartz, to amethyst or red jasper. Some specimens are perfect pointed crystals, others banded agates which somehow recall elaborate confectionery. I last saw the latter on the way to the Moroccan desert, where every bend in the road across the High Atlas comes with a small boy waving agate geodes.* When the curator, Alan Hart, jangled his special keys to let me in on these secrets, I noticed one of the students edging over to have a peep in the drawer. I could read his expression, at once furtive and riveted: "What! Yet more riches!" Alan showed me one mineral that could not be displayed at all to common view. Proustite has a blood-red colour that fades when exposed to light—a shy creature, indeed. It was not named after the novelist Marcel Proust, though one feels it should have been. These famous, well-formed crystals mined from Copiapo, Chile, some of which would almost cover your hand, have an unreal quality, as if they did not quite belong on this Earth at all. I suppose that since they are a compound of silver, arsenic and sulphur they must also be very poisonous. They thus combine a lethal but hidden beauty; they are of the Earth, but somehow also unearthly. Wordsworth was no friend of geologists, whom he regarded

*As if nature were not generous enough with colours, some of the specimens waved by the boys have been dyed rather lurid and unnatural scarlets and purples, colours unmatched by natural quartz. Caveat emptor!

Edward Heron-Allen, polymath, novelist, palaeontologist and historian of the violin

as dull enumerators of facts, but he did write a line that seems quite appropriate to gemstones: "True beauty lies in deep retreats."

Alan Hart then unlocked the drawer housing a special amethyst, safe from causing more trouble at last. The catalogue describes it thus (BM Register 1944, 1): "Quartz (var. Amethyst), faceted, oval (3.5 × 2.5 cm), mounted in silver ring in form of snake, one of which bears two scarabs of amethystine quartz, the other a T in silver, engraved. Locality unknown. Mrs. Mair Jones of London by presentation, January 28, 1944." Mrs. Mair Jones was the daughter of Edward Heron-Allen, one of the great Museum benefactors until his death in 1943, whose collection of an estimated 25,000,000 specimens formed the backbone for studies of single-celled Foraminifera in the Palaeontology Department. Heron-Allen was an extraordinary polymath, a skilled violinmaker and Persian linguist as well as a world authority on "forams." He published everything from novels (*Kisses of Fate,* 1888), translations (*The Ruba'iyat of Omar Khayyam,* 1898), poems (*The Ballades of a Blasé Man,* 1891), history (*Selsey Bill, Historic and Prehistoric,* 1911) and bibliography (*De Fidiculis Bibliographia, Being an Attempt Towards a Bibliography of the Violin,* 1890–94) to natural history (*Barnacles in Nature and Myth,* 1928), not to mention shadier material on what would now be called alternative beliefs (*A Manual of Cheirosophy,* 1885). One thinks of Dryden's lines: "A man so various that he seemed to be / Not one, but all mankind's epitome." Mrs. Mair Jones included a letter from her father that explains all about the curse of the amethyst. Yellowed with age now, it still lies with the specimen in the locked drawer:

> To—whomsoever shall be the future possessor of the Amethyst, these lines are addressed in mourning before he, or she, shall assume the responsibility for owning it.
>
> This stone is trebly accursed and is stained with the blood, and the dishonour of everyone who has ever owned it. It was looted from the treasure of the Temple of the God Indra at Cawnpore during the Indian Mutiny in 1855 and brought to this country by Colonel W. Ferris of the Bengal Cavalry. From the day he possessed it he was unfortunate, and lost both health and money. His son who had it after his death, suffered the most persistent ill-fortune till I accepted the stone from him in 1890. He had given it once to a friend, but the friend shortly afterwards committed suicide and left it back to him by will. From the moment I had it, misfortunes attacked me until I had it bound round with a double headed snake that had been a finger ring of Heydon the Astrologer, looped up with zodiacal plaques and neutralized between Heydon's Magic Tau and two amethyst scaraboei of Queen Hatasu's period, bought from Der-el-Bahari (Thebes). It remained thus quietly until 1902, though not only I, but my wife, Professor Ross, W. H. Rider and Mrs Hadden frequently saw in my library the Hindu Yoga, who haunts the stone trying to get it back. He sits on his heels in a corner of the room, digging in the floor with his hands, as if searching for it. In 1902, under protest I gave it to a friend, who

was thereupon overwhelmed with every possible disaster. On my return from Egypt in 1903 I found she had returned it to me, and after another great misfortune had fallen on me I threw it into the Regent's Canal. Three months afterwards it was brought back to me by a Wardour St dealer who had bought it from a dredger. Then I gave it to a friend who was a singer, at her earnest wish. The next time she tried to sing her voice was dead and gone and she has never sung since. I feel that it is exerting a baleful influence over my new born daughter so I am now packing it in seven boxes and depositing it at my bankers, with directions that it is not to see the light again until I have been dead thirty three years. Whoever shall then open it, shall first read this warning, and then do as he pleases with the jewel. My advice to him or her is to cast it into the sea. I am forbidden by the Rosicrucian Oath to do this, or I would have done it long ago.

(Signed) Edward Heron-Allen
October, 1904

I think we get a fair picture of the Heron-Allen style from this letter. The amethyst (see colour plate 15) is an ordinary-looking stone to have had such a "baleful" history. The curses that lie on several famous diamonds might be construed as a way of discouraging thieves, but a humble amethyst would not be worth that kind of trouble. And it really did wait thirty-three years before finding its way to a Museum drawer, which hardly suggests a hoax. I confess to experiencing a measure of nervousness when I handled the jewel, and I am sure that it was pure coincidence that my back seized up most painfully on the following day. There have been requests to wear the amethyst at Museum parties, but Alan Hart keeps it safely out of harm's way under lock and key. I am told that nobody has yet reported a Hindu Yoga scrabbling at the cupboard to try to retrieve the accursed amethyst and return it to Cawnpore.

There are *some* jewels that are just too valuable to be placed out in the galleries. These are kept in a very large green safe in an office behind a door that opens to a special key. I had better not tell you exactly where the door is. Hidden away, of course, are diamonds. When I was shown

The Big Hole diamond pipe at Kimberley, South Africa, source of many giant gems

the contents of the safe, it did produce a little thrill to hold a large cut diamond, of a size that might interest a seriously rich film star. To a mineralogist, however, an imperfect diamond may well be more interesting than a flawless piece of "ice." Minute inclusions within the body of the diamond can reveal much about its conditions of formation deep within the Earth. Diamonds on the surface of the Earth are strangers from a strange world, hijacked upwards to Garrard's and Fabergé. Tiny bubbles, so small as to be hardly visible under a lens, can be analysed by techniques like Raman microspectography, which uses a laser beam to excite the constituents of the bubble into revealing their spectral properties. No harm comes to the diamond. Diamonds have a simple crystal form. In the hidden collection there is a diamond octahedron the size of a cherry—the octahedron is its uncut natural shape—still emerging from its bed of yellow ground. This was an historic find from 1872 in South Africa, at the beginning of the exploitation of the diamond pipes that founded the fortunes of the De Beer family, who subsequently provided the Natural History Museum with a Director, and who still dominate the market today. Weathered yellow ground preceded the fresher blue ground as the pipes were mined ever deeper, and both are a legacy of the deep event immortalized by a profound duct that runs towards the centre of the Earth. Seventy-five per cent of diamonds were formed in an event about three billion years ago. Scientists are still arguing why this should be, but it means that the impurities in diamonds are potentially one of the best ways to learn about our planet at an early stage of development. For example, large-scale ion microprobes have investigated the sulphur isotopes in tiny flecks included in diamonds that are effectively "fossils" preserved from the early Earth. The results show that at this early stage in the Earth's history the sulphur from the diamonds was likely to have been derived from the atmosphere. Diamonds are a mineralogist's—and not just a girl's—best friend.

They are also a thief's greatest temptation. The Natural History Museum mounted a blockbuster show on diamonds in 2005, bringing together a selection of famous stones. Rare yellow- or blue-hued diamonds could be seen alongside scintillating giants, each of which doubtless carried a special curse. Security was an important consideration; but it was still evidently inadequate, for the show had to close early on 23 November at a loss. Staff were informed that "reliable information" had been obtained by the police that a bold jewel heist was at an advanced stage of planning, and that the greatest robbery in the history of the world might happen at any minute. For several weeks one had to walk past guards carrying sub-machine guns to get to the office while the splendid diamonds were sent securely back to their several homes. It was a slightly humiliating experience—especially as the London newspapers ran articles along the lines of "Mr. Big Spooks Museum." The guards had no sense of humour, either, when an employee gestured towards his briefcase and winked. It was rather a relief when the diamonds had gone home. Though what was once the largest diamond in the world still resides at the Natural History Museum . . . Or, to be truthful, the *space* that it once occupied is curated in the Museum. It is an odd story. The Koh-i-noor diamond is not only one of the largest diamonds but could claim to be the one to which the cliché "steeped in history" most dramatically applies. It is not merely steeped, but is marinated, stewed and spiced in history; naturally, it also has a whopping

Recutting the Koh-i-noor diamond in 1852. Considering the value of the stone, the figure on the right seems rather nonchalant.

curse upon it. It was mined in India in Andhra Pradesh in medieval times, and its early history is more legend than fact, but by 1526 it was probably in the possession of Babur, the first Mughal Emperor. Thereafter it changed hands and country repeatedly as spoils of war: it travelled to Persia with Nadir Shah in 1739; and thence to Afghanistan with the warlord Ahmed Shah Abdali in 1747; and then on to the Punjab in 1813, when it was captured in turn by the Maharaja Ranjit Singh. During its Mughal days, it spent time mounted in the famous Peacock Throne. Finally it passed by right into the hands of the greatest Empress of them all, Queen Victoria. It was an attraction at the Great Exhibition of 1851: "the lion of the Exhibition" according to *The Times.* But its drop-like shape did not please Prince Albert, and in 1852 the stone was recut over the course of eight days for the then astonishing sum of £8,000. During this process it shrank from 37.2 grams to 21.6 grams (that is a 42 per cent loss)—and became much more brilliant. The original was lost for ever.

But, tucked away in an attic of the Mineralogy Department, Alan Hart found a cast of the original Koh-i-noor diamond—it still exists, if only as a hole inside a plaster cast. It had been forgotten since 1852, but could there be a better demonstration of the First Law of Museums: *never throw anything away*?

Notes accompanying the virtual specimen show that the cast was prepared under the supervision of Nevil Mervyn Herbert Story-Maskelyne—one of the leading mineralogists of the time—as a record before the recutting of the famous stone. There was a previous tradition of making glass models of renowned jewels, so replication was common practice. Three hundred and fifty of Sir Hans Sloane's glass models are one of the better survivors from his original collection in the British Museum. Story-Maskelyne became Keeper of the department in 1857, and he it was who first built up the meteorite collection, and arranged the public exhibitions under what was then the cutting-edge, chemical classification, which still forms the basis of the display. He was a pioneer in the use of chemical methods in mineralogy, and demanded a laboratory, in spite of the fact that he had virtually no assistance for a number of years. Nonetheless, he somehow found time to keep on his Oxford professorship, become a Member of Parliament and develop into one of the important early photographers. He was almost Heron-Allenesque in his versatility, though without the interest in mumbo-jumbo. Are we lesser people today, or do we expect less of ourselves? I imagine Heron-Allen passing Story-Maskelyne on the stairs. "Ah! My dear Maskelyne, I regret I cannot spare time for persiflage just now, as I have to finish my monograph in time for the violin recital. Perhaps we will meet at the Sufic Poetry Society this evening?" "Would that it were so," Story-Maseklyne might reply. "But I fear I must complete the determination of the latest parcel of minerals from India before I leave for my constituency in Wiltshire by way of the Welsh silver mines, there to glean materials for my photographic plates."

Whatever our deficiencies of character, we do have better scientific instruments these days; modern scanning techniques may even allow us automatically to recreate the original shape of the Koh-i-noor from its cast—so to build a solid from the vacuum. As for the butchered diamond, it can be seen in the Queen Mother's crown in the Tower of London. And there are those who have attributed Prince Albert's early death to the curse. Although it is a strange kind of death: his ghost still haunts South Kensington—from the Museum and monument that carry his name to the attic of an obscure part of the Natural History Museum.

From another box hidden in the safe Alan Hart carefully lifted the La Trobe nugget—717 grams of native gold, and not the usual formless mass, either, but a jumble of crystals, piled together like a gilded cubist sculpture. Charles Joseph La Trobe, Governor of Victoria, sold it to the Museum in 1857. It remains one of the most magnificent examples of native gold ever found, and it is priceless. It feels curiously heavy in the hand, and somehow it is hard to credit that it is the real thing. One expects it to have a label on it saying 100% REAL GOLD. Another piece of native precious metal from the safe may be even more valuable. It feels a little tasteless to mention value while looking at the treasures in the national collections—rather like the man invited to an aristocratic high table whistling through his teeth and opining out loud that the dinner service must be worth a fortune. But the Kongsberg silver wires really would fetch something approaching a quarter of a million pounds on the open market. The famous silver mines near Buskerud in Norway were opened in 1623, and at their height employed several thousand miners. They stopped production in the middle of the last century, so no more specimens will be found. The wires are very odd—they club together to make a spiral of native silver threads rising from a calcite base, slightly blackened in the atmosphere, but silvery withal. The form is somewhat reminiscent of the famous if unbuilt Tatlin tower of the Russian Constructivists, which was to have been built in 1917; if one was less generous, one might say it was more like a weathered bedspring. Whatever the comparison, it is an extraordinary production of nature, which is why it is so valuable. This spiral "sculpture" was acquired in 1886, but belonged originally to the Geological Survey, whose Museum was formerly a separate institution with its entrance on Exhibition Road. The Natural History Museum took over both the Survey minerals and the galleries in which they were displayed in 1985. The former Geological Museum became the Earth Galleries of the NHM, and the silver spiral was added to the collections.

Everyone would like to find his or her own Aladdin's cave, a jewel-

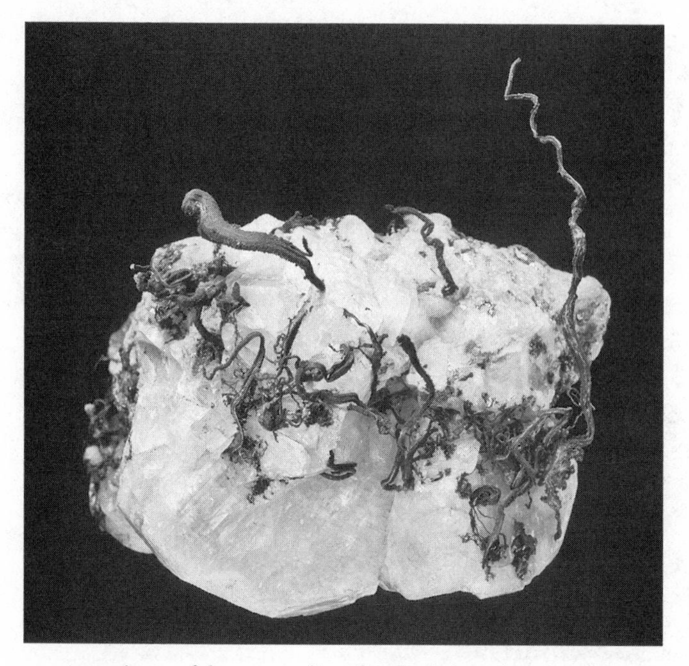

A specimen of the extraordinarily spiral-shaped native silver "wires" from Kongsberg, Norway

clad grotto of the heart's desire. While I was researching this book I gained access to a part of the Museum I had never visited before: not so much an Aladdin's cave as an Aladdin's attic. It was a strange feeling climbing a staircase for the first time in a building where I had spent most of my working life. The Russell Room hides in what would be called the rafters in any regular building. The square columns that pass through the room still carry the effigies of eurypterid fossils—so once upon a time the public would have had access to this corner. But now it feels like the most sequestered recess of a building that is already the apotheosis of nooks and crannies. The room is lined with the polished wooden cabinets and drawers that are the hallmark of the old Museum. Much of it is taken up with the collection of Sir Arthur Russell Bt., who made the greatest ever collection of British mineral specimens. It was received as a bequest by the Museum in 1964. There is an endearing

The doyen of mineral collectors, Arthur Russell, eponym of russellite

black-and-white photograph of the baronet sharing a tandem bicycle with his mother, all Oxford bags and cardigans. He is alleged to have visited every British mine outside the Isle of Man in search of mineral booty. His name has been adopted for the leading society for amateur mineralogists, the Russell Society, and naturally there is a russellite, with its type locality at Castell-am-dinas Mine at St. Colomb Major, in Cornwall, the ancient county that Russell scoured for fine examples. On the top of the cabinet lies his catalogue, immaculately written out in what I would describe as a boyish hand—perfectly formed letters but not joined together. I leafed through it to look for the usual signs of ageing, but in vain; the first entry is as neat as the thousandth.

In the room named after Russell are drawers upon drawers of brilliant minerals—opaque and glittering metallic sulphides, subtle pink and green tourmalines, varieties of spinels and peridots, indeed almost every kind of molecular art that the Earth can produce. Against the wall there is a case of fluorspars from Derbyshire. Chemically, this is just calcium fluoride—but this simple formula provides a rich range of lilacs, and apple greens, or banded displays alternating between colourless and coloured layers, or perfect cubes tumbling like spilled confectionery, making a kind of lithified Turkish delight. Fluorspar is an ancient worked stone, used even by the dynastic Egyptians, but no finer specimens have been collected than those from Derbyshire, on the "backbone of England." In one corner of the Russell Room there are strange metallic-looking "wheels" of a mineral called bournonite purchased in 1879, and today worth in excess of a hundred thousand pounds per sample. They originated from the Herodsfoot Mine in Cornwall. The Lostwithiel mineralogist Richard Talling—after whom tallingite is named—tried to collect the mineral from the mine but was rebuffed by the tin miners. He covertly bought 51 per cent of the shares in the mine, and when he returned to the mine was able to put the miners on to the extraction of strange mineral species instead of their usual ores. In another corner of the room is the type specimen of the element Niobium (formerly Columbium) given to Sloane in 1754 by John Winthrop and used by Charles Hatchett to discover this rare element, which he formally announced to the Royal Society on 26 November 1801.

Science, treasure, rarity, beauty, scholarship: this hidden gallery made me understand again the heterogeneous attraction of Museum life. Nowhere else could a link with the Mughal emperors be relevant to what happens deep beneath the surface of the Earth; nowhere else would the fanatical collecting of a toffish Russell become a long-term resource for mineral genesis; nowhere else could rummaging in an attic reveal an archive of the Prince Regent. From the Russell Room I looked out on to the Victoria and Albert Museum across the other side of Exhibition Road. The prospect might suggest imperial nonsense and "pomp and circumstance," a slightly ridiculous inheritance from the nineteenth century when the Sun never set on the British Empire. But South Kensington has become transformed by time and usage into something that is more than just the "BM" and the "V&A," a monument to a Britain that no longer exists. The collections are there to inform and inspire the whole world, and not just a small corner of it. I am not much of a post-colonialist, and I don't necessarily admire the principles on which the collections were made. But I do understand the primacy of collections as a record of the world, both human and natural. There is more to collections than the golden rule about never throwing things away. There is inherent value in having people who "know their stuff." The apparently esoteric can suddenly illuminate unsuspected areas of knowledge. Those who have devoted their lives to collections—obdurate people, odd people, admirable people—actually make a museum what it is and should be.

8

Noah's Ark in Kensington

Recently, I travelled back in time. Ollie Crimmen, the fish curator, had somehow got hold of an old BBC1 *Horizon* programme about the Natural History Museum broadcast on 7 September 1970. It carried the title of this chapter. The whole programme had a faded feel to it, as if it had been recorded in about 1902. All the scientific staff were wearing V-neck jumpers and ties and spoke in clipped, upper-middle-class tones. One could imagine that their pipes had only just been tapped out. The blue-uniformed warders looked as if they were doing the job after a stint on E-wing of a prison for violent offenders. The exhibitions they supervised really *were* still those inherited from 1902, comprising lines of things in glass cases, with technical labels in small type. The Director—Sir Frank Claringbull—was filmed in a classic book-lined study facing the camera as straight as if it were an interrogator, and talking too fast. There were also some of the people who have appeared in this book. Here was the famous whale man Peter Purves, explaining elegantly how he aged whales from the deposits laid down in their ears, as he wandered through the skeletons of a dozen cetaceans that were formerly stashed in the basement. Purves displayed no sign of insobriety at this stage in his career. Then there was Peter Whitehead, fish man and roué, explaining very clearly why taxonomy mattered, his beard still black, the bags under his eyes at an early stage in their ontogeny. Tony Sutcliffe, fossil mammal man, was on film looking just the same as when he retired twenty-five years later—I realized he had always looked about sixty years old. Miriam Rothschild, the Trustee's Trustee, handsome in a mannish way, wearing a curious kind of cape and backed by ranks of leather-bound volumes, was speaking of collections in a way that showed how deeply she cared about them—*her* BM collections, the value of which was beyond dispute. The kids on the galleries seemed much more timeless, fidgeting and tucking into their sandwiches. I realized that memory is a great deceiver. It muddles things up, it does not square with time, and it does not move along in the way one expects. The documentary evidence of this film proved how it really was. I was alarmed to find that there were people in the film that I remembered not at all: the bat man was somebody I may have never met, or have forgotten, and therefore will always be excluded from my personal museum. The way that he delicately examined a pickled bat's ears suggested that he might have been interesting to know; that he, too, would have had stories to tell. In a strange way this demonstration of the limitations of memory proves the importance of collections in museums. They defy time; they transcend what any one scholar might make of them; they are outside our own little personal histories.

The official histories of museums tend to be dominated by the few for whom archives are compiled and maintained, and most particularly Directors. This is not so different from the way that the history of England used to be portrayed as the history of kings and queens, or that of the United States of America by the doings and ponderings of presidents. It is certainly a convenient way of keeping a chronology in mind—a kind of temporal mnemonic—but it hardly paints a true picture of what happens among the people: acts and edicts are the least part of history. I have been concerned with what goes on behind the scenes among the practising scientists. I have not referred very much to the politics at the top of the organizational hierarchy. As in William T. Stearn's history, I could have recounted the scientific story Keeper by Keeper, but that, too, is a history of successive governments rather than a history of fallible and interesting practitioners. However, it would be remiss of me not to write a little about the running of the Museum by the boss—what is today termed "governance," a word that to me sounds like a cross between "government" and "performance" and automatically carries a whiff of admonishment. The Directors of the Natural History Museum have gradually dwindled in public prominence over the last century and a half. Earlier Directors included figures that could take their place on the world stage of savants; more recent occupiers of the post are competent people put there to run things, not to influence the perception of the natural world in the human one. When Richard Owen agitated for the foundation of a separate Natural History Museum, he could directly commune with the royal family, at a time when that family really counted for something. Nobody disputed that he was also one of the greatest comparative anatomists of the day. Even his independent views concerning evolutionary theory might be attributed to some kind of comparative resentment over Darwin's intellectual ascendancy—there was room for only one top dog. There were not many people in the nineteenth century who could command influence as effectively as Sir Richard.*

His successor, Sir William Flower, was Director between 1884 and 1898. A convinced evolutionist and friend of Thomas Henry Huxley, he it was who placed evolutionary theory at the centre of the exhibitions. He commissioned a large marble statue of Darwin by T. E. Boehm at a cost of £2,000, a serious sum then. It was unveiled on 9 June 1889 by the Prince of Wales, in the presence of Admirals J. Sullivan and A. Mellersh, both of whom had been on the expedition of the *Beagle*. Owen's statue had to wait until 1900. Flower is not as familiar a name as Darwin or Owen, but he established the principle of exhibiting informative specimens while protecting the study specimens behind the scenes in the scientific collections. He was a devoted public servant, whose health suffered in the over-conscientious pursuit of his duty; ill health forced him into resignation. He is also the only Director with a surname appropriate to the job.

Edwin Ray Lankester, who followed Flower, was another great figure in public science. He had already been Linacre Professor in the

*To be accurate, I should say that Owen never had the formal title of Director—he was officially entitled Superintendent.

Zoology Department at Oxford, so he was top of the academic tree before he even crossed the threshold of the Museum. His *Textbook of Zoology* (1909) became a standard work. But he was also well enough known to the public to appear in cartoons in *Punch* heavily riding upon the back of the okapi (*Okapia johnstoni*), which he had made known to science and the press in 1901. This odd relative of the giraffe was first known from bands of skin collected in the Congo a year or so previously and thought to be a new kind of zebra. Lankester, through the "Africa hand" Sir Harry Johnston, later acquired a whole skin and two skulls. I like a description of Lankester in the journal *Candid Friend* of 1901: "His own head is shaped like a benevolent biscuit-tin and is packed as full of knowledge as other people's eggs are full of meat . . . the only thing that moves the excellent and bulky biologist to unmitigated wrath is a real idiot." He was a naturally commanding figure, tall and paunchy, with a deep, loud voice that could be heard along the length of the longest gallery. Like many clever people he did not tolerate fools gladly, and he had a most capacious definition of "fool." This led him into outbursts and confrontations that made his time at the top a difficult one. He had rows with the Trustees. He had rows with the BM. At the end of the nineteenth century the Principal Librarian at Bloomsbury, Edward Maunde Thompson, was one notch up in the hierarchy from the South Kensington Director, something that Ray Lankester found irksome. He was reprimanded by Sir Edward for not informing him of the dates when he was away on leave, at which point the memoranda started to fly. The seriousness with which both men took the definition of the "pecking order" suggests two large egos unwilling to budge, and as often happens in such situations, the conflict escalated until the Trustees eventually had to reinforce the status quo. Now the disappointed Natural History Museum Director sought to take on the Trustees in turn. Lankester was well connected at the Royal Society, from which he would one day receive the highest honour, the Copley Medal, which was awarded to Stephen Hawking in 2007; Albert Einstein and Max Planck were previous winners. Through behind-the-scenes lobbying at his "club" of top scientists, he sought to persuade some influential Fellows to write a memorandum publicly complaining of the inadequate way in which the Museum had been governed by its

Edwin Ray Lankester, Museum Director, riding on the okapi he described (from *Punch*, 12 November 1902)

Trustees. His machinations came to nothing. By 1904 his relations with the Trustees had deteriorated to the point where they appointed a subcommittee to report on the running of the Natural History Museum—which it did, in no uncertain terms. An extract from the report reads:

> When Sir Richard Owen was appointed head of the Natural History Museum he appears to have regarded the post as being in the nature of a reward for scientific eminence, while administration and superintendence were to occupy a secondary position. Sir W. Flower exercised a closer and more systematic superintendence over the Museum than had been the practice of Sir R. Owen. Prof. Lankester appears to take the view that his duties and functions are such as were undertaken by Sir R. Owen, rather than those that were fulfilled by Sir W. Flower . . . We fully recognize the great value of the scientific researches prosecuted by the Director, but at the same time we

> are strongly of the opinion that in the interests of the Museum the duties as laid down by the Statutes should be strictly carried out in future . . . in conformity with the practice usual in the case of other Civil Servants.

Edwin Ray Lankester's rather large knuckles had been thoroughly rapped, and his days were numbered. He was retired at sixty on 31 December 1907, but not before fighting the Trustees' rights to dispose of his services every step of the way, including a huffy letter to *The Times.* He was, incidentally, one of nine people at Karl Marx's funeral. Although his history might suggest that he was what would now be termed an elitist, his prolific essay writing for the general reader shows the contrary: an unpatronizing clarity, and a capacity to charm. I read *Diversions of a Naturalist* (1913) with pleasure, and I maintain that Lankester might be mentioned in the same breath as J. B. S. Haldane as an occasional writer. No doubt his dedication to science and his intemperate and unqualified belief in his own rectitude led to his difficulties. But nobody could deny that he was a man of real intellectual substance and an ambassador for biology.

However, the Trustees had voiced their opinion: administration should come before charisma, efficient systems before science, and "usual Civil Service practice" should obtain throughout. There should be lots of nicely presented reports and memoranda. A Museum Secretary would help the smooth running of things, and indeed one Charles Fagan had proved to be indispensable in this role for decades in the earlier half of the twentieth century. The Director should be the oil between the cogs that makes the whole machine run without a glitch. However, there was at least one more Director of global stature*—although I should perhaps rather say "globular stature" since Sir Gavin de Beer was both short and stout. He was Director from 1950 to 1960. He was born with such a large number of silver spoons in his mouth that he must have found eating a challenge. He spent his early years in France,

*One could also make a case for Tate Regan (Director, 1927–38), whose distinguished scientific career preceded his elevation, and who saved the British nation from the musk rat, and appointed women to the permanent staff for the first time.

Sir Gavin de Beer, Director in the mid-twentieth century—Sir Cumference

which conditioned him to be multi-lingual, a polyglot polymath. While he was President of the Fifteenth International Zoological Congress in London in 1958, he prided himself on addressing all the European delegates in their own tongue. He was most extraordinarily clever, and very aware of the fact. He wrote on evolutionary theory, particularly with regard to embryology; he penned a biography of Darwin; he described *Archaeopteryx*, the famous early fossil bird, and he wrote volumes of history in his spare time. His bibliography extends into hundreds of articles, reviews and books. He arrived and left every day in his Rolls-Royce, immaculately besuited, and it was common knowledge that he had to perch atop a pile of cushions to get a fair view of where he was going.

Everything had to be just so for Sir Gavin: a flunkey had the small lift awaiting his arrival in the morning, his desk was laid out in a particular way. His loyal and efficient Museum Secretary, Thomas Wooddisse, took care of much of what the Trustees expected, leaving Sir Gavin time and opportunity to play the Great Man. Many honours were loaded upon him. He did have a pompous grandeur that the shop-floor staff were pleased to mock. Edmund Launert tells me that the Director was described as "Sir Cumference." In the Botany Department he was known as "Volvox," which is a small green alga forming spherical colonies in perpetual motion in water. Many of the stories from this time relate to his vainglory, and the puncturing thereof. My favourite tale involves sausages. One of the oddities of oral history is that you get similar stories from several sources, but the details change. There is no holotype for a story, as there is for a butterfly species. Nobody seems to know who the original sausage actor was, although there are several people who attest to the story's veracity. In the basement, during the 1950s, it was the custom of members of staff to cook up breakfast on one of the Bunsen burners that were plumbed into the offices there. According to Phil Palmer, one of the sausage cookers was called off to do something else during the middle of a breakfast fry; by the time he got back he was horrified to find the sausages ablaze in a haze of fat. Without thinking too hard about it, he recalled the drinking fountain in the main hall—just the place to put out a modest fire. He took off up the stairs with his flaming frying pan only to run into Sir Gavin in the company of King George VI on an impromptu early-morning visit. As the *Punch* cartoons used to say: collapse of stout party. I do hope the story is true, and not the product of wishful thinking.

Vainglorious or not, Sir Gavin Rylands de Beer was certainly a scientist of global significance, the last of such stature to hold the post of Director. Sir Terence Morrison-Scott, who succeeded him, was a snobbish, smooth civil servant, a worthy but hardly spectacular mammalogist, whose natural habitat was probably an establishment club like the Athenaeum. I am sure that he would have been approved by the same Board of Trustees that reprimanded E. Ray Lankester. Sir Frank Claringbull, who was Director from 1968 to 1976, when the documentary film was made, was a mineralogist by trade, and his main achievement will be seen as the initiation of the modernization of the galleries, which looked so dated on the old *Horizon* programme. Updating the "show" was the most important change in the public perception of the Natural History Museum, and soon was responsible for a vast increase in visitors. During the war years Claringbull had the curious distinction of growing the largest crystal of Trinitrotoluene (TNT) known to mankind, and would suddenly produce it from a large matchbox to discombobulate his visitors. Ron Hedley, Director from 1976 to 1988, was the last to be recruited from the ranks of the scientists. He was one of very few Directors not to get knighted for his services, and one can only speculate why. Possibly he resisted the "business first" philosophy which was being pressed upon all the public services during the Thatcher years. For the shop-floor scientists he was still one of us, and, with the Keepers of the departments still around the table, the central administration continued to have a scientific flavour to it. In the vaults, the work carried on as usual.

Ron Hedley had initiated one political change that proved to be more significant than we could ever have known at the time. The Natural History Museum had for some years had the status of Research Council. Much of government money in science is channelled through the Research Councils; the giant among them is probably the Medical Research Council that supports laboratories up and down the length of Great Britain. The Natural Environment Research Council is another large organization which handles science that is generally closest to that pursued by natural historians. The Natural History Museum was a Lilliput by comparison, and there was a feeling abroad that we passed under the radar as far as inevitable government cuts were concerned. Hedley took us out of the Research Councils and into what was then the Office of Arts and Libraries—where we joined the other national museums like the British Museum, V&A, and the National Gallery. It may have appeared a logical move at the time, because it seemed likely that the Research Councils would be required to make economies yet again—and governments are less prone to snip at flagship cultural establishments because it doesn't go down well with Tory voters. The finances were already difficult enough because entrance charges had been introduced by the Tories, and it was unclear what long-term effect this would have on attendance. Whatever the combination of reasons for the administrative move—and it was effective financially—the result was to end the unique status of the Natural History Museum as a research establishment dressed up in the clothes of a tourist venue. Now we were just one museum among several. "Front of house" was as important as behind the scenes, and science had to pay its way, with no space for slackers or eccentrics or unaccountable presences like the Baron of Worms. It was evidently time for reform.

Reform came in the person of Neil Chalmers, formerly Dean of Science at the Open University, who took over as Director in 1988. He had published a number of papers on monkeys, but I am sure he appealed to the Trustees mainly as an administrator of the new school. It was time to break the link with scientists that had supplied Directors for so long. Scientific distinction simply became less important for the job. I have already described the bloodletting that happened subsequently as many of the staff were "let go." Peter Whitehead's novelistic prophecies began to have a ring of truth about them; there was a certain madness abroad. Perpetually smiling, Chalmers always seemed so pleasant and enthusiastic, rather like a vicar welcoming the parishioners to a car-boot sale, yet people around him had a habit of "disappearing": Clive Bishop, Keeper of Mineralogy and Deputy Director, went early and unhappily; Julian Legg proved to be the last Museum Secretary, having been dismissed for some irregularities which were never explained in detail; he was followed by Dr. Lawrence Mound, Keeper of Entomology. It began to seem rather dangerous to be too close to the new Director. The shop-floor scientists looked on, buffeted and bemused. A new post of Director for Science was introduced—a *capo dei capi* for science in the Museum, and the first one appointed was John Peake. But on the central management team the Director of Science replaced what had previously been *all* the Keepers of the five departments—so now he was a lone voice for science, outnumbered by the newly minted Directors of all the other functions of the modern museum: human resources, public engagement, finance, estates and so on. The day of the scientist at the centre of Natural History Museum life was over. By the time Chalmers, by then Sir Neil, left in 2004, a new Director, Michael Dixon, could be appointed who had spent virtually all of his life in administration rather than at the scientific coalface. I cannot see a return to the magisterial, if slightly comical and egocentric, style of a Sir Gavin.

And there will certainly be a new logo. One of the first things that happened when we said goodbye to the British Museum (Natural His-

The logo for the Natural History Museum introduced by Dr. Chalmers, and ubiquitous for the "brand." It soon became known as the "zebra's bum."

tory) was the appearance of a new logo—for the Natural History Museum, finally and officially named as such. It was decreed that the logo should appear on all the stationery and the posh envelopes. I do not suppose that many of us had heard of branding at the time—but this was our *brand.* It was designed by Wolf Ohlins, a company which had designed a fair number of logos and brands, down to the last detail of typeface. One got to recognize their designs: they were a kind of calligraphic shorthand crafted to convey the spirit of the product. The Liberal Democrat Party got a skeletal phoenix; the World Traveller class of British Airways got wavy lines. The Natural History Museum got a tree-like form, which might have suggested an evolutionary tree—or even perhaps a real tree—or perhaps a metaphysical quality of "treeness." Whatever the intention, the shop floor soon christened it the "zebra's bum" and that is what stuck. It got on our personal visiting cards and we all grew quite fond of it. A new carpet, adorned with thousands of little zebras' bums, appeared along the gallery floor that led to the Palaeontology Department. When Dixon was appointed he decided that the Museum's logo was "a little tired," and out went the zebra's bum and in came a large capital "N" to replace it. Up came the carpet. I could just imagine Sir Gavin spinning around in his grave like a corpulent top. In fact, I find myself spinning around just a little in sympathy.

At the same time, important changes were happening everywhere else in the organization. The Museum took over its own freehold, and became responsible for the upkeep of the famous building. Scientists were freed up to apply for grants from the Research Councils now that we were just another organization competing on the open market. These grants were always very difficult to get, and would get still harder to win during the last decade of the twentieth century as competition from the universities increased. So far in the twenty-first century nothing much has changed. Furthermore, grants became more and more important for the mere survival of research programmes that had hitherto been core funded from the central pot. Management loved grants because they brought in "overheads"—which meant money for other parts of the Museum. But many fundamental kinds of taxonomic research have not proved attractive to granting bodies so that research is increasingly tailored to win grants. The requirement to bring in as much money as possible led to the expansion of the shopping area in the galleries and the proliferation of all manner of trinkets for sale to the kids who come to see the dinosaurs. Small, fluffy tyrannosaurs will growl at you and sing the theme song from *The Sound of Music.* The Museum increased the number of books and catalogues it published and expected them to make a profit. The area given over to children and education increased mightily. All these activities required more staff, so that the proportion of scientists in the total staff roster continued to shrink. We needed to make more of a fuss about everything we did—including science—in order to increase our media presence, and the Public Relations staff increased commensurately. Nice women with smart suits and lipstick and bright smiles attempted to bring out the scruffy old scientists from their hidden redoubts. Their elbow patches were confiscated. Corporate culture had arrived, and sent that old sepia world packing, and not before time, most people would have said. The Natural History Museum had become an Attraction! Roll up! Roll up! And nobody could regret too much the passing of those galleries full of ranks of glass-topped cases that would have been recognizable to E. Ray Lankester. Everything was somehow so much brighter out there now.

I was involved with the beginnings of all this. The first of the new galleries was the hall of human biology—"making an exhibition of ourselves" as the label eventually read. As a very young scientist, I was in charge of the first steering group for this trail-blazer for the new exhibition schemes, a show all about human biology, perception, the brain, development, genetics and even society. It could not have been more of a departure from the old public galleries. My team made out the first brief for the exhibition, full of bright ideas, not all of them practical. I read everything from Richard Gregory on visual perception and the brain to Jane Jacobs on the growth of cities; it probably did me a lot of good. By the time the exhibition actually opened in 1977, most of the original outline had disappeared. By then, an able ex-palaeontologist, Dr. Roger Miles, had been placed in charge of the modernization of the exhibitions. He was a rather formidable figure: serious-minded and with distinctive convictions about the educational value of the new displays. He approached the exhibitions with the purposefulness he had previously applied to the description of the Devonian fossil fish from Gogo, Australia. Some people did not approve of his high-mindedness, which they associated with a Nonconformist strictness. As one of them remarked to me at the time: "There is Methodism in his madness!" But if Miles had a didactic side, I do not believe it mattered a jot when it came to applying it to the galleries. The Human Biology galleries were a great success. They managed to be both fun and informative, with plenty of knobs to turn and audiovisual displays. It is easy to forget how revolutionary these exhibits were at the time, and even thirty years on they do not seem out of date. I will give one example of their effect. I had noticed that not long after the exhibition opened there seemed to be a large number of young Spanish girls going into the exhibition. This was a time when Spanish au pairs were a common sight in London. I followed them (in the most academic way) and noticed that they beat a path to one particular exhibit, which dealt with the growth of the foetus, inside a womb-like room, illustrated by a riveting film of this vital part of the reproductive cycle. One or two of the girls were making notes. I suddenly realized that this was important to them for one simple reason: this was probably the only sex education they had ever received.

The status of women within the Natural History Museum itself has changed fundamentally from the early days. From the first there were remarkable female scientists who made important contributions to the collections, as well as those talented botanical artists who contributed artistically to floras. Beatrix Potter was a talented and accurate painter from life, although she did not think much of the employees at the Museum: "a pretty dull lot" as she remarked. Then there were those female adventurers who took off to dangerous parts of the world to secure specimens for the collections. Thanks to a biography by Karolyn Schindler, we can now appreciate the courage of Dorothea Bate, who was a pioneering palaeontologist and collector of mammal fossils from the Mediterranean region. Starting in the early years of the twentieth century, when she was still a young slip of a thing, she explored Cyprus, Crete, Majorca and Menorca in search of new genera and species of fossil animals. What are now popular holiday destinations were then quite wild areas in which brigands could cause trouble, and roads were so poor that access to sites was either by boat or by exhausting treks over dangerous tracks. Nor was there such a thing as sensible field gear for a lady in those early days. Dorothea Bate was undeterred by the obstacles, even when, as she wrote to the Keeper of Geology, Arthur Smith Woodward, finding fossils was "like searching for the proverbial needle in the bundle of hay."

In the caves known as Cuevas de los Colombs on Majorca, she was rewarded by finding bones of a very peculiar animal that stood only about forty-five centimetres to the shoulder, which she was later to name *Myotragus* (Greek: "mouse-goat"). The caves lie on the edge of the sea underneath some formidable cliffs; one can easily imagine what it must have been like to try to gain landfall when the Mediterranean coughed up one of its famous bad storms—and even in calm weather the cave system was several hours by boat from the nearest harbour. *Myotragus* was remarkable in lacking the usual six lower incisor teeth so typical of ruminants but having instead just two, and these huge and continuously growing, like the gnawing teeth of a rat. It was an impossible animal! News of the discovery excited great curiosity back at the Natural History Museum, for nothing like the beast had been known before. Dorothea was fortunate that she was encouraged by two successive Keepers of Geology, Henry Woodward and Arthur Smith Woodward, who not only valued her work but paid her enough for the specimens she collected to allow further expeditions. The new discover-

The skull of the curious extinct mammal *Myotragus* from the Balearic Islands, discovered by the redoubtable Dorothea Bate

ies were exhibited in the Museum as mounted skeletons and the public flocked to see them. Dorothea also found pygmy elephants in the mountains of Cyprus. She was largely responsible for initiating a story about the Mediterranean islands during the Cenozoic Era that is still having flesh added to its bare bones. The genesis of *Myotragus* happened when a small species of goat migrated into the Balearic Islands from the Iberian Peninsula at a time of low sea level about six million years ago. At this so-called Messinian time, the Mediterranean Sea was sealed at the Straits of Gibraltar, and nearly dried out. Thick, natural salt deposits were formed in a deep basin where the sea had once been; harsh times, indeed—but for a few animals this crisis was also an opportunity to reach new land. When sea levels rose once again, the "goat" became isolated on its new island home, and over a hundred millennia or so it evolved its extraordinary form while adapting to life on a rocky island: it probably also became a superb climber. Isolation of other islands led to further endemic species—in fact, the phenomenon of "dwarfing" in island species is known elsewhere, so the pygmy elephant is perhaps not so surprising.

Dorothea's fearless exploration of the islands was a measure of the new confidence of women as scientists—and probably not just a local phenomenon, when one remembers that Marie Curie won her Nobel Prize in 1903, a symbol that women might win glittering awards as well as the vote. But, as for remuneration from the Museum, Dorothea Bate had to live most of her life on what we would today call "soft money"—hand-outs of one kind or another. At the outset her expeditions were financed, or specimens were purchased, as a result of the excavations. There is no doubt from the correspondence between Dorothea and her male Museum sponsors that she was taken seriously as a scientific worker—the language is collegiate, not patronizing. However, when she finally came to join the Museum staff she was not properly established, but paid on a piecework system by the number of specimens she prepared.

There were, if it is possible, even more redoubtable and indomitable women collectors. Lucy Evelyn Cheesman was probably the most remarkable. She was born in the Victorian era in 1881, and survived into that of the Beatles, dying in 1969. As a skilled entomologist she travelled through the South Seas, the Galapagos, the Moluccas, collecting indefatigably as she went. In 1924 she left the St. George's Expedition when it arrived in Tahiti and continued on her own. In the same year she was made the first woman curator of the London Zoo. Between 1929 and 1955 she went on a series of expeditions more or less by herself to Papua New Guinea, Indonesia and the South-West Pacific—she was truly intrepid. The thickest jungles did not seem to deter her nor did negotiating with unlettered natives. She published dozens of scientific papers on her discoveries, mostly on insects. But it may be characteristic of the time in which she was working that some scientists seemed to be unaware of her gender. A man named Riley described a species of tomato, based upon specimens Cheesman had collected from the Galapagos Islands. He named it (*Solanum*) *cheesmanii*—which is a masculine ending. Presumably he thought that only a man in shorts and pith helmet could have got to such a remote spot. Fortunately, there is justice in nomenclature: the same species name has lately been corrected as *Solanum cheesmaniae,* a feminine ending acknowledging the gender of the discoverer. There is a photograph taken in the field of Evelyn Cheesman when she was of a certain age; thin as a lath and with a set

The fearless entomologist Lucy Evelyn Cheesman in the field

jaw, eyes shielded under a hat; one would know at once that this was a woman not to be trifled with. New Guinea tribesmen would fall silent under her imperious stare. But she was no termagant either: she wrote well and humorously about her adventures in *Six-Legged Snakes in New Guinea* (1949) and *Things Worth While* (1957). But neither is there any sense of a woman in subjugation to a male hegemony: the males wouldn't have dared.

When women were finally admitted on to the scientific staff under the directorship of Charles Tate Regan, they soon became some of the best-known scientists in the Museum. Because no quarter was given to their sex, they might have been described as "blue stockings" by the less enfranchised of their sisters; for the most part I would rather describe them as dedicated. When I came to the Museum to work on fossils, I was introduced to one of the greatest women scientists, Sidnie Manton.

She was the high priestess of the arthropods, and over a long research life studied many of them, from sea spiders to lobsters to moths. She had a sister, Irene, who was at least as famous as a botanist, and the Manton sisters remain unique as the only sisters elected to the Royal Society. When I was appointed to my job, Sidnie was married to John Harding, Keeper of Zoology, but she was not formally on the staff—although she spent most of her time working in the Museum. The studies that made her famous included very detailed descriptions of the embryology of crustaceans, and even today workers turn to these meticulous scientific papers. She went on to discover how the arthropods worked, how their muscles lifted and moved their jointed legs from inside, and how different kinds of arthropods—millipedes and spiders, for example—had very different gaits. She treated her subjects like complex pieces of machinery. I often think of Sidnie Manton when I watch *Star Wars,* in which aliens use gigantic metallic arthropods as steeds, all struts and spindly beams and ball-and-socket joints. Because she found fundamental differences between different groups of arthropods, she came to believe that arthropods had evolved more than once from different ancestors. This idea has fallen into disfavour since molecular evidence has indicated that they did indeed ultimately descend from a single ancestral "proto-arthropod" about a billion years ago. A few scientists have been rather patronizing about Sidnie Manton, as if this one topic was all she had touched, but this is to neglect the enormous factual contribution she made to our knowledge of the most numerous animals on Earth. She had a major stroke in her last decade, one that left her almost completely paralysed. But she went on to write her great summary book about her life's work, *Arthropoda* (1978), composing at a typewriter with a finger that still worked. There was something peculiarly tragic about seeing one who had done so much to improve our understanding of scuttling animals ending her days so immobilized, but it was impossible not be inspired by her unquenchable determination in the face of misfortune. Her work on the crustaceans continues unabated with a well-known team of researchers in the Darwin building. With fish stocks now diminishing, the importance of prawns and crabs and their myriad relatives to the feeding of humans is bound to increase—but much needs to be learned about their life histories, parasites and habitats before sustainable systems can be worked out. What goes on in the back rooms of museums has a vital contribution to make to understanding the complexities of marine ecosystems.

Theya Molleson was a close colleague of Kenneth Oakley, who exposed the Piltdown fraud, and she pioneered the study of diseases as they leave their mark on human bones. She did sterling work in the 1980s on the human remains interred in the crypt of Christ Church, Spitalfields, where there was enough biography associated with the bones to investigate such things as the effects of diet, society and wealth on a section of historical London. The Spitalfields Project was also a good example of collaboration between very disparate bodies: the Greater London Council, the Institute of Archaeology, the BM, the Natural History Museum and so on. Over in the Zoology Department Juliet Clutton-Brock was the authority on the domestication of animals, and wrote an entertaining book about the subject. She left the Natural History Museum as a result of a late imbroglio from the Chalmers "night of the long knives." When notable scientists retire, they are sometimes given the honour of a *Festschrift,* a special volume of learned papers on their speciality written by their colleagues. It is a rather touching homage. Nobody except Juliet could have gladly received the volume presented to her in 1993 bearing the title *Skeletons in Her Cupboard.* Dr. Ailsa M. Clark is a world expert on starfish, and rather formidable; she was also irreplaceable after she retired . . . And for every woman scientist there were, and are, three or four indispensable female curators labouring away on the collections . . . and beyond them again there are female volunteers who still give their services for nothing. It is not an exaggeration to say that nowadays science would fall apart without the contribution of women, and it seems extraordinary that the secret world behind the scenes at the Natural History Museum should once have been an almost entirely male preserve.

The old Museum may have been hidebound by petty rules, but the staff's security of tenure meant that members of staff were free to be naughty. I am told that just before the war years there was an illicit still inside the model of the blue whale—the same huge replica that survives on exhibition in the public galleries. Indeed, it was given a spring clean in 2007. Inside it is hollow, making it an excellent hiding place—a trapdoor in the stomach allowed access. Stuart Stammwitz, who became head of exhibitions, was allegedly responsible. He was the son of Percy Stammwitz, who for many years was the Museum taxidermist. Together they were one of the few examples of the Museum becoming a family business. I should add that the trapdoor was later sealed over, but a telephone directory and some small change remain hidden inside. For a short while, the workmen who built the whale could have felt what it was like to be Jonah. Then there was the case of the dodo. The bird gallery is the last of the old-style galleries: stuffed with stuffed specimens. Like the minerals gallery, it has been allowed to survive from the old days with which this chapter opened. It is a parade of labelled ornithological examples, family by family—with the added delights of a specially detailed display case on eggs, showing the biggest and the smallest. A somewhat apologetic and in my view unnecessary notice explains that this is a deliberately antique gallery, left behind to show how it used to be done. Halfway along the gallery is a fine dodo, the famous extinct bird, all decked out in whitish feathers. It is, of course, a bogus bird, since no perfect specimens survive in the collections. In fact, it was largely based upon a painting, which may or may not have been painted from life. The feathers are stuck on. The talented model-maker Arthur Hayward was asked to make the life-size replica in the early 1950s—and swan feathers were just what was needed to make the thing convincing. The trouble is that Thames swans belong to the Guild of Lightermen or to the Queen. One is simply not allowed to go and grab a swan. This didn't deter Barney Newman, another of the Museum's distinguished topers. He and an accomplice went down to Hammersmith Bridge and grabbed a large cygnet from under the bridge, where nobody could see what they were doing. It was stuffed into a bag and thence into the back of an unmarked van. One version of the story has it that they were stopped by a policeman while speeding back to the Museum, and that Barney had to do creative coughing every time the bagged bird struggled in the back of the van. For all that the Museum was once a stuffy, hierarchical place, decades ago the staff knew how to subvert the rules. If they were "established" civil servants

The dodo (*Raphus cucullatus*)—
a reconstruction using swan feathers

they had to do something very bad *indeed* to get the sack. Members of staff now may have more freedom, but it comes with less security.

The real wrongdoers have been persons associated with the Natural History Museum, but not on the staff—conmen who abused a privileged position to dupe the curators and researchers. The worst by far was Colonel Richard Meinertzhagen, who died at the ripe age of ninety in 1967. Adventurer, soldier, ornithologist, big-game hunter and spy—he had the kind of biography that makes one blink in disbelief.* He was as striking in person as his life history would suggest: a natural soldier, erect in posture, chiselled features under a short beard, an imperious

*Like many people who constructed much of their own legend, he wrote prolifically to support the tale. Perhaps the most prescient title was *Diary of a Black Sheep* (1964), but one should notice *The Life of a Boy* (1947), *Kenya Diary 1902–6* (1957), *Middle East Diary 1917–56* (1960) and *Army Diary 1899–1926* (1964). There have been several adulatory biographies that take him on his own assessment, but it has become clear that fabrication played an important part in the "diaries," too.

and charismatic mien coupled with a penetrating intelligence. He was given to wearing distinctive clothes: one of my elderly colleagues remembers playing table tennis with Meinertzhagen, who was dressed in a black cloak that flew around him "like the very devil." He was aristocratic and well connected—he even went for childhood walks with Charles Darwin. His passion was birds, and nobody doubts that he was highly skilled as an ornithologist. His postings in Africa and the Middle East allowed him access to rare species, and since he was also an able hunter his collection of skins grew rapidly. In 1954 he presented his collection to the Natural History Museum—an enormous gift of twenty thousand skins. He also had an interest in lice, and presented another collection of about half a million of them mounted on glass slides and in spirit bottles. The Trustees approved Meinertzhagen as an Honorary Associate, a title given to few amateurs, as a measure of their gratitude for the receipt of his collection. No doubt the fact that he was a "toff" did him no harm at all in those snobbish times. Thereafter, he could stride around the Museum as if he owned the place. I regret that I am just too young to have met him, for he was one of those rare people who impress themselves on others within a few seconds. Such people should, perhaps, be fenced off from their fellows from an early age since little good ever comes of such charisma.

Colonel Richard Meinertzhagen survived with his reputation undiminished and died a hero. Seventeen years after his death, his scientific history began to unravel. There were many early indications of untrustworthiness that had been ignored in the aftermath of his generous gift. He had been under suspicion of purloining specimens some years beforehand—there had been a partially successful attempt to ban him for a time from the "bird room." When the extent of his duplicity was revealed at last it cast a cloud over much of his published research. He had stolen specimens from other museums and incorporated them into his collection, and then claimed to have recovered the specimens from new localities. In astonishing examples of sang-froid, he even took specimens from the Natural History Museum's own collections, relabelled them, and then presented them back to the institution with a flamboyant bow. Thanks to careful research by Robert Prys-Jones, the original status of specimens was reconstructed. For example, consider

The labels tell the story of Meinertzhagen's deception
on an owlet specimen.

the Blyth's kingfisher (*Alcedo hercules*), which Meinertzhagen claimed from two specimens from Burma. It transpired that both specimens had been stolen—one from the Whitehead Collection at the Natural History Museum, the other from the Owston Collection of the American Museum of Natural History in New York—and both were really from Hainan Island, China, where they should have been.

As Prys-Jones and others dug into Meinertzhagen's collection, more and more bogus examples were discovered. It began to seem appropriate that one of the animals he really *did* discover was an African giant hog (*Hylocheorus meinertzhageni*). The problem with the whole collection is that many specimens are perfectly genuine, but careful research work is needed to establish which ones. The type specimens of the Afghan snowfinch (*Montifringilla theresae*) are undoubtedly authentic, to take just one example. This species is named for entomologist Theresa Clay, thirty-three years Meinertzhagen's junior, who was his "companion" latterly, and who had catalogued the insect collections he had presented to the Museum. They lived next door to one another at 17 and 18 Kensington Gardens with a passage connecting them. The more one finds out about Meinertzhagen, the more dangerous he seems. It was said that he killed a native assistant while in India, an incident that was covered up as a death from plague. Theresa Clay, I should add, was a charming woman who I am sure had no conception of her lover's dark history. Others had seen his character more clearly from the first. Lawrence of Arabia said of him in *Seven Pillars of Wisdom* (1926) that "he was willing to harness evil to the chariot of Good." Meinertzhagen had been an effective spy in Palestine during the First World War, and the business of spies is duplicity. It is hard to construe his motivation for zoological fakery as anything other than a love of mischief and confabulation for its own sake. He could have enjoyed an unsullied reputation had he wished to. He was what novels of his time would have described as a cad and a bounder—and the most dangerous of his kind, one gifted with a luminous personality.

Another figure broke the rule of trustworthiness without which scientific research cannot operate. Arthur Kingsbury died the year after Meinertzhagen. He was a solicitor and amateur mineralogist who during the middle decades of the last century built up a reputation for finding astounding specimens in the field, especially in localities in Cornwall. It was all most extraordinary: a group of mineral enthusiasts would be grubbing around on a spoil heap for some interesting species, when Kingsbury would suddenly wave around a stunning specimen of the object of desire—quite the best ever to have been discovered from the locality. He dressed distinctively in old-fashioned, colonial gear in the field, complete with knee breeches, and became something of a legend. His abilities were attributed to a miraculous sixth sense about where the good specimen lay hidden—and nobody asked awkward questions. Well, scientists don't make things up—it's against the rules. After he died his widow sold the collection to the Natural History Museum. The first misgivings about authenticity began to emerge in the 1980s. Collectors who had not been in thrall to the Kingsbury mystique began to complain that they could not duplicate his results from certain localities in spite of their most diligent efforts. Eventually George Ryback was employed to undertake a detailed investigation. The earlier specimens in the Kingsbury Collection were respectable, having been mostly acquired from other collections. But many of those that had astonished his fellow collectors in the field were "plants"—they were from well-known localities elsewhere in the world, and Kingsbury must have had them secreted on his person, only to whip them out at the right dramatic moment. Detailed mineralogical studies showed that a gold specimen (Reg. No. 1965, 83) said to come from Porthcurnick Beach, near Portscatho in Cornwall, actually came from Kanowna, Kalgoorlie, Western Australia. The occurrence of native gold atop manganese oxide crystals is thoroughly characteristic of the latter locality. Then there was a green fibrous malachite specimen that Kingsbury claimed to have found at the Driggith Mine in Cumbria, but which was identifiable as having come from Zellerfeld in the Harz Mountains of Germany. This list went on and on.

Oddly enough, I believe I can understand Kingsbury's motivation rather better than I can Meinertzhagen's. When I was a student I had one day on a field trip during which I genuinely had a "magic hammer": every locality yielded its treasures instantly to my most cursory tap. My fellow students were half admiring, half envious. The don in charge took notice of me, eventually nodding significantly at me when we arrived at a new locality and indicating that he would not have long to wait for me to find a diagnostic fossil. It was a good feeling (it didn't last) and I can understand how somebody might want to get a reputation for omnipresent luck. However, the whole scientific endeavour relies on truthfulness. Scientists are not supposed to make mistakes, though almost every scientist will have done so once in their career. Mistakes can be corrected, admitted to and even forgiven. But deliberately to mislead is the ultimate sin in science. It is the fact that both Kingsbury and Meinertzhagen died unexposed that really rankles. It also remains true that such duplicity is rare, and that the integrity of the collections is preserved by frankly admitting to the bad hats and mountebanks.

There were greater threats to the collections than the activities of charismatic fraudsters: two world wars. During the First World War the main threat to the collections seems to have come from the government itself, which several times tried to purloin the space for its own purposes, notably to house huge numbers of clerks in 1918. The Natural History Museum never entirely closed, and provided solace and entertainment for convalescent troops; it continued to have about half a million visitors a year throughout that war. William Stearn discovered an

Specimens had to move to safety during the Second World War; a large snake moves to a storage facility deep under Surrey.

entertaining faux pas by the Speaker of the House of Commons. In 1916, during one of the attempts to close the museums (Bloomsbury, the Tate Gallery, the Science Museum, the National Portrait Gallery and the Wallace Collection were also included in the plan), the politician spoke in a disparaging manner of "deciphering hieroglyphs and cataloguing microlepidoptera," the implication being that all real men were out there in the trenches doing useful stuff. He could not have picked a worse example. Hermetically sealed tins of army biscuits destined for troops in all corners of the Empire proved to be full of maggots when they were opened. An entomologist at the Natural History Museum, John Durrant, was called in by the War Office—and the maggots proved to include the larvae of three species of flour moths, all of them belonging to Microlepidoptera. The Speaker was also one of the Principal Trustees of the Natural History Museum, so he should have known better. The Museum contributed to the war effort in a dozen ways rang-

At 4:30 a.m. on 9 September 1940 two incendiaries and an oil bomb hit the roof of the Botany Department; the damage to the collections took years to repair. Stored seeds germinated.

ing from the treatment of body vermin to advice on which wood to use for aircraft. As in every other institution in Britain, most of the younger men saw active service and many did not return to their benches.

The Second World War brought much more serious threats to the Museum. If one wanders up Exhibition Road to the entrance to the Earth Galleries, it is easy to see where lumps were gouged out of the limestone blocks in the wall: a bomb fell in the road, and could easily have demolished the adjacent buildings. These holes are the last visible evidence of the bombing of this part of London. The war in the air made things much more dangerous for the type collections, and a decision was made to evacuate most of them, beginning in August 1939. Various stately homes deep in the countryside became temporary quarters to butterflies or molluscs from the national collections, and the aristocratic owners welcomed them enthusiastically, for the most part, as preferable to having a whole lot of troops billeted on them. The public galleries were closed, and then they were open, and then closed again as circumstances changed. The first closure happened on 29 August 1939; by 3 February 1940 the Museum was opening on Saturday and Sunday, and, later that same month, daily; by 29 May—closed again. It must have been confusing for the visitors, in spite of announcements in the newspapers. The archives show that many of the scientists who remained in the Museum did work for the Red Cross or as auxiliary firemen. But the Natural History Museum was a conspicuous target, and life was exciting from time to time. An oil-bomb fire started in the Botany Gallery on 9 September 1940 at 4:30 a.m. and caused much damage. Some seeds of the silk tree, *Albizia julibrissin,* collected from China by Sir George Staunton in 1793, got an unexpected soaking, and proceeded to germinate. This was the first of a number of hits and near misses—there were three more before the year's end.

At the eastern end of the old building, where the palaeontology wing now stands, a series of concrete bunkers were built. They are still there under the new building—I have always known them as the War Rooms. Now they form a low basement housing some of the anthropology collections, boxes of human skulls and bones, a dark and oppressive place. When they tried to blow the rooms up many years later, the exceptional thickness of the concrete made it an impossible task, so the War Rooms were incorporated into the new building instead. Airborne attacks continued sporadically until 11 July 1944, when a flying bomb in the Cromwell Road did severe damage to the western galleries and the central towers. The Museum had reopened to the public again on 1 August 1942, but closed once more on 6 July 1944, the day after the first flying bombs landed in Queen's Gate on the western side of the building. At the end of the war one could say that the Natural History

Museum had come out of it better than might have been expected. Waterhouse's extravagant creatures were still on their perches. The longest-paid member of staff on record was remunerated as a result of the damage to the Botany collections: Arthur Hales was brought back to sort out the specimens he knew so well, reconstructing their curation history from surviving scraps of writing. He eventually retired, the job done, at the age of seventy-three.

The Second World War years are recorded in a house magazine called the *Tin Hat*. The first number was produced on 30 September 1939. It is illustrated with charming, if crude, cartoons, often in the deft abbreviated style of Fougasse. The tone is best described by using the wartime word "chipper"—cheery in an eye-rolling way. I particularly like the section on overheard remarks appearing under the heading "The Brighter Side." They tended to record humour from moments of extreme tension, as when somebody remarked after the bomb damage to the bird room that he had never seen so many birds killed with a single shot. According to the instructions, contributions to "The Brighter Side" "should be sent to Messrs Claxton or Smith. They should be brief." Another bird-room example will give you the idea. "The shattered Bird Gallery! Broken glass, torn blinds, blasted doors and window frames, dust and grit everywhere. And at the gallery entrance lying conspicuously on the floor, a large printed label 'BIRD MIGRATION.'" We learn about the Air Raid Protection shelter, and the suggestion that it may have had bugs. In June 1944 the first number of another magazine, 6323, was issued, which published news of staff fighting in the forces all over the world. There is something rather eloquent about these pages, maybe having something to do with the old-fashioned typewriter on which it appears to have been rapidly knocked out. If it had been done these days on a computer it would be slicker but somehow more impersonal. Peter Purves appeared to me in these pages for the first time, complaining of mud everywhere and fungi rotting everything in the jungles of the Far East. 6323 was the telephone number of the Natural History Museum at the time, as it still was when I joined the staff several decades after the war had finished. Looking at these fading pages, I sense a strong connection with all those curators and collectors, scientists and plant pressers, an unbroken string of scholars united by the

Cheery under fire: the cover of an edition of the wartime house journal *Tin Hat*

collections, enduring through peace and war, small champions of order in a world where chaos is always a possibility.

At one time it must have seemed possible to grasp the totality of the world's biodiversity, to discover the entirety of what the Reverend Gilbert White would have called "the system." In the days of Britain's imperial greatness such optimism informed the initiation of great expeditions to map the unknown and collect its life. On 7 December

1872 HMS *Challenger* set off around the world under Captain John Nares to find out about the oceans and the life they contained. This was an expedition in the tradition of the famous *Beagle* voyage of Fitzroy and Darwin, but with a particular scientific focus on the least-known two-thirds of the globe. Britain was the leading maritime nation, and a voyage like this was then a top research priority. Richard Corfield has redescribed the journey that founded the science of oceanography in *The Silent Landscape.* The ship returned in 1876, having taken samples of water, sea floor and organisms at 362 different sites. Among the many discoveries was the recognition of the Mid-Atlantic Ridge, and the Challenger Deep in the Marianas Trench, not to mention the realization that life—in rich variety—could exist deep in the oceans in the realm of eternal darkness. *Challenger* was possibly the first vessel to be kitted out fully with purpose-built laboratories in order to allow conservation of specimens as soon as they were collected. The obligation to publish the results occupied the talents of several employees of the Natural History Museum for years. Under the editorship of John Murray the *Challenger* report, or *The Report of the Scientific Results of the Exploratory Voyage of the HMS Challenger During the Years 1873–1876,* was one of the great achievements of scientific investigation, running to some fifty volumes published in the decade 1885–95, each volume weighing in at about the same size as the family Bible. More than four thousand species new to science were named in one or another of these works, and many of their type specimens still reside on the shelves in the Darwin Centre. I have looked at a series of crabs sadly contemplating me from inside their glass jars, *Challenger* having granted them this strange kind of immortality. If only enough expeditions could be mounted it might be possible to catalogue the whole world, and bring it back to the shelves of a great museum for perpetuity! And indeed there were many more collecting trips, but perhaps nothing on so grand a scale as HMS *Challenger.* Brave individuals voracious for specimens, such as Evelyn Cheesman, or even Colonel Meinertzhagen, added mightily to the collections and the coverage of the remoter corners of the world, but these collecting trips were tasks that knew no end. By a cruel transmogrification of Parkinson's Law—"work expands to fill the time available for its completion"—biodiversity apparently expanded to match the best

HMS *Challenger,* the foundation of scientific oceanography

endeavours of experts to get to know it. Possibly the last tailor-made expedition run from the Museum was the African entomological shindig in the converted truck during the 1970s with Peter Hammond and Dick Vane-Wright.

Meanwhile, the world was changing, and paying little attention to the Natural History Museum. In the early days it would have been assumed that an unthreatened series of global ecologies would be there for sampling in perpetuity. The rapid disappearance of marsupial mammals from Australia following the introduction of cats and foxes showed how vulnerable species could be to mankind's interference. This lesson was reinforced when knowledge of the riches of island faunas and floras like those found on Hawaii was coupled with an awareness of their fragility and vulnerability to extinction. During the twentieth century, destruction of whole habitats continued to gather pace, particularly in India and the Far East, and the systematic mission was more and more coloured by such phrases as "before it's too late." The taxonomist was beginning to be both christener and obituarist. Taxonomic collections and skills were reborn in mitigation of this new and harsher world. If dense concentrations of species could be recognized in particular places and habitats, such diversity "hot spots" could be more readily targeted for protection and conservation. The Worldmap Project was started in 1988 at the Natural History Museum, using computer mapping of species based on Geographic Information Systems (GIS) technology. The entomologist Dick Vane-Wright and the botanist and computer virtuoso Chris Humphries began the project and laid down the ground rules for recording species distributions, and Worldmap is now fully operational under the amiable guidance of Paul Williams. Digitization of collections and their occurrences, as well as many other kinds of biological records, allowed any area of the world to be interrogated for its taxonomic richness. To take one example, it is obvious from the biodiversity maps that the Malay Peninsula fully deserves protection efforts to be concentrated upon it. The identification of unknown species from this area should be a priority, but even before that stage is reached conservation of threatened habitats should ensure that such species do not become extinct before they can be identified. That is the theory, at least—but human rapacity often outstrips good intentions, as has happened for many years in Indonesia, where habitat destruction of rainforest is continuing at a dizzying rate. Museums have no political power, but they do have the possibility of influencing the political process. This is a complete change from their role in the early days of collecting and hoarding the world to one of using the collections as an archive for a changing world. This role is not merely scientifically important, but it is also a cultural necessity. I believe the organizers of the *Challenger* expedition would have approved. But it is a long road from the quiet vaults of a museum to staying the hand of an illegal logger.

The change in the scientific role of the systematist and taxonomist has been profound over the same period. I have touched upon this several times previously when describing the detailed work of individual scientists, but it bears repeating that the original role of the systematist as cataloguer of life's diversity has been supplemented and in many cases replaced by a role as investigator of its interrelationships. Many of the younger generation of scientists are interested in methods of working out phylogeny—evolutionary trees—and are not necessarily wedded to "their" group of organisms. We have come across the use of molecular evidence in unscrambling the tree of life, as used in everything from nematodes and winkles to mushrooms. It's a whole new catalogue of data useful in understanding the multifarious strands that weave together the biological world, and provides a different way of looking at nature from the old "hairs on legs" of insects, or pistils and stamens of flowers. Such is the complexity of DNA that computer methods for handling the evidence it provides are indispensable, so there is a now a career as an analyst for the kind of scientist who is more at home with programming than with looking at weeds or bugs. The green screen figures more prominently than the microscope in their offices, and conversation with their colleagues revolves around the virtues or deficiencies of a new piece of software rather than the discovery of a new species of butterfly. I must admit to being one of the old-fashioned kind for whom the collection remains paramount. Sometimes I think I am already halfway into an historical collection myself. Already I can imagine some future curator peering at an old departmental photograph and wondering out loud: "What an odd-looking fellow . . . I wonder what he knew about?"

There is a persuasion I have come across among the scientists and curators that the right way to die is slumped in front of the microscope at an extremely old age. In the right hand the quill pen will just have scratched out the last species description of a huge and complex group of organisms. The old boy or girl will have a vague smile upon that wrinkled but deeply distinguished face: a job well done. Then another curator should come along and, one hopes, stick a label on the recently deceased and incorporate him or her into the collections, another fine specimen of *Homo taxonomicus.* Or maybe deposit his cadaver in the demestarium to be stripped down by the beetles. Several of my fellow travellers on the commuter train from Henley on Thames have expressed astonishment that I should still go into the Natural History Museum to work *for nothing* after my official retirement in 2006. Of course, the project is bigger than mere money. When I first joined the Museum there were still a few ancient figures staggering up that grand cathedral entrance, like the beetle expert "Tiger" Tams, who wore a dark suit and a white wing collar, and looked as if he might have walked straight out of the staff photograph taken at the time of E. Ray Lankester. The brachiopod worker Ellis Owen is eighty-four as I write, and still comes in once or twice a week just to defy our notions of how people are supposed to look older as the years pile on. Even while the cultural shifts I have described in this chapter repeatedly change the face of the Museum, these brave people continue to work to their own agenda, before they themselves are inevitably curated by time. The duty towards discovery is not something that can be lightly cast aside. These people feel it at a very profound level, and it is not connected to financial reward, and only occasionally to public recognition. Their motivation is an unquenchable instinct to find things out and to make these discoveries known to others. Their duty is directed towards an inventory of the biosphere, which now needs their services more than ever before. It is probably one of the better manifestations of what it is to be human.

9

House of the Muses

At last we emerge from one of the polished doors that lead down into the secret world of the scientists—out again into the galleries where a ten-year-old child pauses momentarily, thunderstruck by seeing the *Tyrannosaurus rex* that has haunted his imagination. Most of the pressing crowds will have no idea of the hidden life this book has explored, the secret world of scholarship and collections, of type specimens and curators, of busy analytical machines and biomedical research. I suspect that a far greater proportion of the visitors will know that the world is in trouble, that "the times they are a-changin'," that familiar animals and plants are becoming scarce, and that in some rather unspecified way something should be done about the rainforests and the oceans. The connection between the unprecedented environmental transformations of modern times and the Natural History Museum will not be obvious, but readers of this book will now know something about why such museums matter so much as an archive of nature. For the public, the exhibitions are there to give a show, and to inform. During my working life they have changed from being worthy and didactic to become "attractions"—a choice among many available in London. Those who decry such changes should remember that when the okapi was first displayed people would travel especially to London to see a single mounted animal. This has always been a function of museums—they are the place to show off a worthwhile spectacle. Adrienne Mayor has demonstrated that even in classical times princes in the Mediterranean regions collected fossil mammal bones and displayed them to their subjects as evidence of the battles between the gods and the giants. They were the Roman antecedents of Dorothea Bate.

If the source of the word "museum" is a house of the muses, then the original museum might have harboured all the arts and sciences, corresponding to the nine muses. The first building to carry the name was probably the university in Alexandria about 300 B.C. Only one of the muses, Urania, Muse of Astronomy—she who is portrayed in a mural in Herculaneum pointing at the heavens with a staff—would have personified the scientific endeavour remotely in the modern sense. From the early days of the British Museum the display of classical archaeology and art was an important part of the function of this new public space. This was a nod of recognition of the new civilization towards the old, a kind of acknowledgement of mutually shared culture.

Naturally, museums came to resemble their classical ancestors, as shown so blatantly at be-columned Bloomsbury, and a dozen other similar establishments around the world. When the Natural History Museum eventually broke away from the BM, the architectural model was also broken. Waterhouse's extravaganza was something new: part cathedral, part fantasy, for now natural history did not have to follow the classical model. But the convention of exhibiting rows of objects accompanied by scholarly notes persisted for far longer. Early pictures of the interior of the Natural History Museum portray a veritable crowd of natural history specimens. If you want to see a museum that has hardly changed from this model, go to the Natural History Museum in the middle of Dublin, which has been frozen in time for decades—there is something wonderful about its profusion of skeletons and stuffed animals and glass models of every kind of invertebrate. When our own exhibitions entered their modern phase in the last few decades, the contrast between the Natural History and Science Museums and the other great art galleries could not have been more profound. The art or antiquarian gallery was still a shrine to the displayed object, whereas the other kinds of museums included models and "push buttons," quizzes, computer games and thrills like simulated earthquakes. I well recall when the Earth Galleries in the Natural History Museum were modernized how specimens were *put away* back behind the scenes—down to the vaults from which they had once emerged: less was regarded as more, providing it was dramatically lit and accompanied by signage to satisfy all age groups. In the old museums more was evidently more, and it was impossible to over-stuff a gallery.

By a kind of post-modern backwards somersault that would keep a cultural analyst like Jean Baudrillard amused for hours, the other muses have sprung back into the galleries. I have seen ballets performed to illustrate the genetic code—and so welcome back to the Muse Terpsichore! Since the process of inheritance is full of movement, what could be more appropriate? Andy Goldsworthy has produced exceptional sculptures based upon leaves or mosses or twigs that sit quite naturally in the galleries, although they have little connection with the didactic function of the displays there: artists may take what they will, and use what they want in unexpected ways. "Sci-art" of this broad kind has become something that charities will fund and that artists will seek out. I was "shadowed" by a fine poet called Deryn Rees-Jones who wrote interesting verse arising out of the experience; but she took from my trilobites things that I had never seen before, as befits a poet.

It seems that the future may yet lie in the past: the museum as a total cultural experience. I have mentioned that a few of the old-style galleries remain in the Natural History Museum—rather better lit than in former times, but still at heart a parade of diversity. The ichthyosaurs and plesiosaurs form a kind of fossil wall, a dolphin-like phalanx suddenly ossified, and then stilled for ever. The bird gallery lies beyond them, lined with real, if stuffed, birds, and one fake dodo. There is even the oldest exhibit in the Museum hidden away there, a collection of hummingbirds in a splendid case, perched or mounted in mid-flight—it is more art than science, and it still attracts admirers. Neither of these galleries has much in the way of modern explanatory accoutrements: yet I do not think I am imagining that both still seem to be rather popular in the early twenty-first century. Children evidently like to see the real thing, at least those youngsters below a certain age—maybe ten or twelve. Older children tend to gravitate to touch screens and audiovisuals. I wonder if we are seeing a return to the object in the science-based museum. Since any visitor can go to a film like *Jurassic Park* and see dinosaurs reawakened more graphically than any museum could emulate, maybe a museum should be the place to have an encounter with the bony truth. Maybe some children have overdosed on simulations on their computers at home and just want to see something solid—a fact of life. The dinosaur gallery has attempted a compromise between reconstruction and actual material that most visitors find satisfactory. I conclude that the way a museum displays its collections and its knowledge changes and changes again, and that this is how it should be. Occasionally, the changes are part of a project by government, as when we were encouraged to think of social inclusiveness—though I doubt if exhibits were ever prepared with an eye to social *exclusiveness.* No social baggage is carried by displays of dinosaurs or butterflies. The natural world belongs to everyone regardless of race, colour and creed—and in the same democratic spirit we are all responsible for its survival.

Life behind the scenes is also changing. The most serious challenge facing the specialist today is how to continue to pursue the systematic mission, how to write and publish the large monographs which summarize chunks of the natural world for posterity. Monolithic individual achievements are harder and harder to complete in the marketplace of the twenty-first century, because the modern museum demands that their researchers should be money raisers, entrepreneurs and computer experts as well as knowing about "their" beasts. In my view the description, naming and discussion of the evolution of organisms should still be at the heart of their work, because unrivalled collections and libraries allow museum scientists to do this kind of thing better than anyone else. This is also the work that lasts. Ernest Rutherford did science a tremendous disservice when he made his famous remark about all science being physics—and everything else being stamp collecting. It is on a par with George Bernard Shaw's aphorism, "those who can, do; those who can't, teach," in glibly undermining the self-worth of devoted and mostly under-rewarded professionals. In a sense, taxonomy and systematics *are* stamp collecting, if that means laying out definitive identifications in catalogues where the user can identify a specimen to hand. The catalogue happens to be the description of what four billion

Richard Owen entertains to dinner some of the animals with which he has been scientifically concerned. Owen published in the monographs of the Palaeontographical Society, which still produces standard works on British fossils regardless of the vagaries of fashion.

years of life's history has achieved, and its contents are a measure of the health of the planet. Isn't that enough?

Those who share Rutherford's view would contrast this kind of scholarly work with "hypothesis-driven research"—and they would regard the latter as the real business. When I began to research this book I soon found examples of hypotheses generated by members of staff at the Natural History Museum that had once been taken seriously but have now been consigned to the lumber-room of history. W. D. Lang worked in the Geology Department on the small, mostly marine, mat-forming colonial animals known as bryozoans. We have met them previously. About the time of the First World War several theories sought

W. D. Lang, Keeper of Palaeontology 1928–38, and devotee of orthogenesis

to question the importance of natural selection as the driver of evolution; I should emphasize that neither Lang nor his contemporaries questioned that evolution had happened. Lang favoured an idea termed orthogenesis—the notion that there were certain routes on which evolution could embark, and once an organism had taken such a course its descendant species were destined to follow the trail even if it led to eventual extinction. In the case of Cretaceous cribrimorph bryozoans, this meant that individuals of the colony, or zooids, in their limy cases became progressively calcified and "walled up" until they reached the point where they were obliged to seal themselves in. They built their own tombs. It was the palaeontological counterpart of Edgar Allan Poe's claustrophobic tales of being incarcerated or buried alive. Lang referred to lineages of bryozoan species as being "doomed to extinction" with all the drama of a finger-wagging sermonizer. Maybe it is no

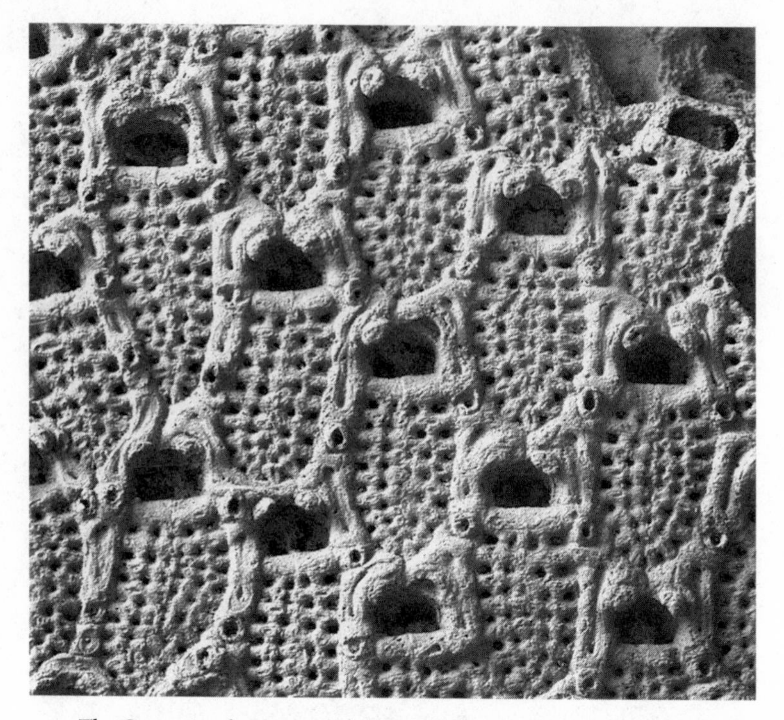

The Cretaceous bryozoan *Pelmatopora.* Lang saw the cases in which the zooids live as becoming inevitably more and more "walled in," eventually resulting in the extinction of the lineage.

coincidence that Lang was a serious Christian: indeed, he published an article entitled "Human Origin and Christian Doctrine" in *Nature* in 1935. By the same mechanism, the Cretaceous bryozoans' terrestrial contemporaries, the dinosaurs, got larger and larger, and both the lumbering monsters and the lowly marine mats came to a similar sticky end at the same time.

Paul Taylor, who is Lang's successor in the Museum in the twenty-first century, has shown how his hypothesis blinded him to recognizing that bryozoans related to the Cretaceous ones did, in fact, survive into younger strata—how could they, if they had already followed the path of doom? Lang's theories were shared by a fair number of scientists, and for a short while might even have been regarded as "cutting edge." It did not take long before the critics pointed out the sheer silliness of such scenarios—even Lang's colleagues in the palaeontology rooms at the British Museum (Natural History) were unsparing. F. A. Bather wrote of orthogenesis in 1920 that "a race acquires the lime habit or the drink habit, and, casting off all restraint, rushes with accelerated velocity down the easy slope to perdition." What survives of Lang's work now is almost entirely his systematics, the names and descriptions of fossils he published, while his "hypothesis-driven research" is but a curiosity. The fate of his theory did no harm to his career prospects, and Lang followed on from his critic, the level-headed echinoderm authority Bather, as Keeper of Geology in 1928—a post he held for a decade before retiring to Charmouth to a house appropriately if toe-curlingly called Lias Lea.*

A devotion to bold theories is not confined to distant days. A year or two before I joined the staff of the Natural History Museum Dick Jefferies published the first of his many scientific papers on some odd, ancient fossils called carpoids that are found in rocks of Cambrian to Devonian age, that is, from about 520 to 364 million years ago. I have discovered new species of these peculiar animals in Ordovician strata, and there is no mistaking them when the rock is split open to reveal one lying inside because many of them are curiously irregular looking—Dick always compared these animals to the shape of a boot. They had a covering of calcite plates, like many marine animals, but the plates have a particular structure which is typical of the Phylum Echinodermata—"spiny-skinned" animals: sea urchins, starfish and their allies. So for many years they were assigned to that group of animals without much comment, although it was acknowledged that they were oddballs in the phylum. Dick Jefferies had other ideas. He believed he recognized features on these strange creatures that were fundamentally similar to those on primitive chordates—the group of animals that includes vertebrates like fishes, frogs and ourselves. Over many years he described

*Some readers may not know that the Blue Lias is a Jurassic formation that crops out on the southern coast of England near Charmouth and is famous for fossils, which Lang wrote about in several publications. Lang's theory is similar in some respects to that of L. F. Spath (see p. 88), who used ammonites as his "experimental material."

minutely and named many of his animals, and fitted them into the basal "twigs" of a tree of relationships that included major groups of chordates. Dick's opinions were expressed in a deep voice, punctuated occasionally by a hearty guffaw; he is one of those people who dig deeply into anything that captures their attention—so he knows all about linguistics, and Chaucer, and is fluent in German, which he learned because he needed to read classical German works on embryology. For a while his theories were taken seriously by many of those interested in the deep branches of the history of life. Dick's basso profundo could be guaranteed to ring out across the conference floor if any topic touching on vertebrate ancestry was raised. At the same time molecular evidence confirmed an old idea that vertebrates and echinoderms had descended from a common ancestor perhaps a billion years ago. Dick's ideas were taken up by one of the editors of *Nature,* Henry Gee, who wrote a book, *Before the Backbone,* giving the carpoid hypothesis favourable coverage. Research students came to work in the Natural History Museum on new discoveries of the strange fossils, including some I had found in the Ordovician rocks of Shropshire in the company of my friend from the National Museum of Wales, Bob Owens.

However, throughout the 1990s, support for the Jefferies theory was slowly ebbing away. Several of Dick's students came around once more to the idea that carpoids were a specialized group of echinoderms, more closely related to one another than to the several basal branches of the chordate tree.* They were an interesting side-branch of evolution rather than a seminal one. Some of the evolutionary transformations that the Jefferies theory required began to seem overly complicated. As it got more elaborate, the theory began to buckle under its own weight. When more examples of new species were collected, even the geological record of the fossils seemed to be distinctly out of phase with the supposed relationships on the Jefferies tree of descent. Dick did not take kindly to being deserted by his ex-students. He would mumble darkly about being "stabbed in the back." It was distressing to see the pain this

*I have over-simplified the Jefferies theory here, which later came to include a wider variety of animals, including an obscure but evolutionarily important group known as the hemichordates.

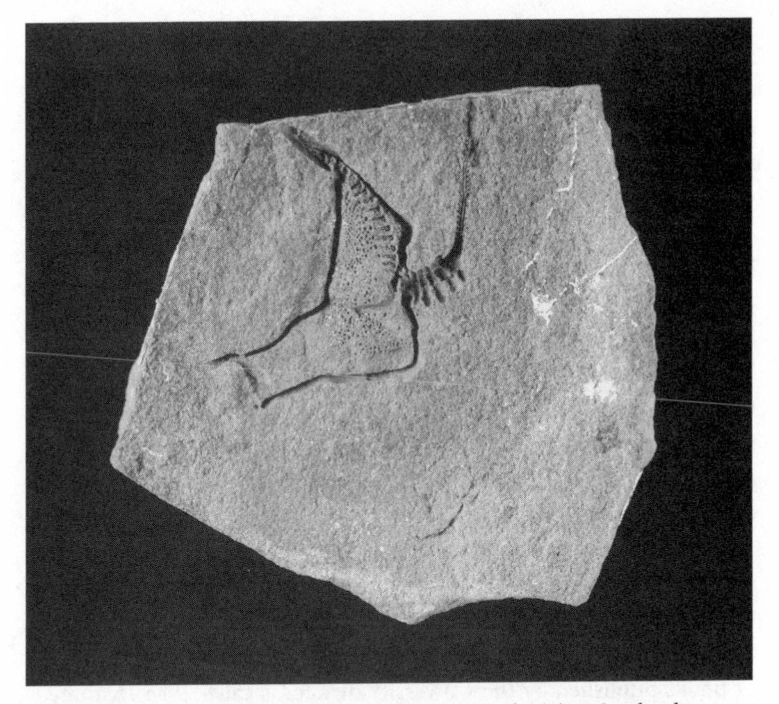

The carpoid *Cothurnocystis elizae* Bather, Ordovician, Scotland. These odd animals have been the centre of a long debate about the origin of vertebrates.

caused to such an admirable and profound scholar. As with so many people in this book, he has continued to work on undaunted after official retirement, still convinced that the world will come around to his view. I hope that the world does just that, but it is progressively unlikely. However, regardless of the grand evolutionary design, all future students of the carpoids will have available Dick Jefferies' names, and meticulous descriptions of these long-extinct organisms. This work will last, even as theory moves on, as theory should. I do not suppose that this will cheer up the good Dr. Jefferies, true though it is. He has a slightly wistful air these days.

I have explained in the previous chapter how the capacity to secure external funding is now regarded as a sine qua non as a measure of success for scientists. Now we come to a crucial paradox. It is very hard to get grants to study organisms and update their taxonomy. From the accounts of current research I have given in this book it will be clear just how varied is the research agenda across the Museum. Some kinds of research—such as that on schistosomiasis, or on insects that damage crops—are obviously commercial. Other fields of investigation, such as that on marine nematodes, might sound arcane at first, but can soon be rallied to the cause of monitoring environmental pollution and the like. It might be hard work to get external funding to study nematodes and greenfly, but it can be done. Mineralogy and mammon have performed a pas de deux ever since the days of Georgius Agricola. The history of mankind always has a certain cachet, and anthropologists come in with a head start in the Palaeontology Department, followed closely by dinosaur experts. Scientists with less obviously appealing areas of expertise begin to have a hard time.

Complaints about failure to receive funding can begin to sound like whingeing, and the list of unsuccessful applicants soon gets to be a long one. A short account of one of my own disappointments might stand in for many others. The definitive summary of fossil biodiversity is a series of books published by the University of Kansas called *The Treatise on Invertebrate Paleontology*—we usually just call it *The Treatise,* in the same way we might talk about the Bible. Most palaeontologists would not argue with the proposition that it is the most generally useful reference in the whole subject. Volumes treat taxonomic groups, such as trilobites or brachiopods, genus by genus, family by family. The original Trilobita volume was published in 1959, since when we have learned much, much more about these fascinating fossils. A revision was planned in several volumes. The first of these finally appeared in 1997, edited by my old Professor, Harry Whittington of the University of Cambridge; it covered only a small fraction of the trilobites. I was then asked to edit the subsequent volumes. With my other commitments it was clear I needed help, which meant giving employment to a bright young post-doctoral assistant. A grant proposal was prepared to this end with *Treatise* completion as the goal and sent to the UK's Natural Environment Research Council. It got the bum's rush; worse, it did not even pass through the first sift by the college of assessors to receive the scrutiny of an independent referee. Presumably, lack of hypotheses being tested was lethal; it was, after all, just a piece of stamp collecting. As a result I have been obliged to resign as editor: goodbye *Treatise.* Researchers more adept than I have managed to obtain funding for "classical" systematics by hitching up with molecular sequencers. For example, they might examine how systematics prosecuted using classical morphological characters match up with phylogenies derived from molecular evidence. We have already seen how Sandy Knapp got support for a definitive study—including molecular work—for the tomatoes and their allies. Recall David Reid and his winkles, or Paul Kenrick and his primitive plants. Many more proposals meet the same fate as did mine—consigned to the funding wastebasket. I know of people who reckon to spend almost half their time trying to raise money, time that might be better spent furthering their knowledge of animals and plants for the benefit of everybody. It is recognized that there has to be some form of selection, but classical systematic work is selected out too early.

However, given skilful legerdemain a systematic study can be wrapped in the finery of hypothesis testing, and the best young systematic biologists have become expert in a kind of creative duplicity. I suppose that I must be identified with those old sepia-coloured scientists with leather elbow patches if I protest that none of this tomfoolery should be necessary; that making known the biological riches of the world both past and present should not require subterfuge. I assert that taxonomy has never been more important now that so many of the world's precious habitats are under threat. Knowing about biodiversity and taking steps to preserve it is no mere luxury—it may prove to be the criterion by which we are judged by our descendants. The pragmatic assertion that we might lose species of service to mankind—as a future source of drugs, say—may be broadly correct, but is not a morally compelling argument for conservation. Suppose, for example, that it could be proved that a rather rare species had nothing of worth for us humans locked away in its genome—would any decent person then set about exterminating it as a worthless organism? I doubt it. A better appeal for the pragmatist might be that we cannot know the consequences of damaging ecosystems that have taken hundreds of millions of years to develop. The precautionary principle is generally a wise one—but it is probably already too late for delicate habitats, especially in the oceans. In any case, the contention that we should know our fellow inhabitants of the home planet seems to me beyond argument. Biologists have done their best to unify respect for biodiversity by initiating such projects as the Tree of Life, wherein hundreds of individual scientists are collaborating to stitch together a map of relationships for all the organisms on Earth. The project reminds me of a collaborative patchwork quilt that was pieced together by a number of West Coast feminist artists in the late 1990s, in that the individual pieces have merit, but the completed whole will be more than the sum of the parts, and will make a political and moral point. On a practical level, some individuals contributing to the Tree of Life dealing with one or another group of organisms are obtaining funding for their work. The memorable label helps.

The basic kind of description and naming that Museum employees have carried out for so long is often called alpha taxonomy, beta taxonomy being systematic analysis and construction of evolutionary "trees." This fundamental if superficially unglamorous science is the most difficult of all to fund, and, as the entomologist Henry Disney said in 2000, continues in "relentless decline" despite lip-service being paid to the importance of taxonomy following the 1992 Earth Summit in Rio de Janeiro. A positive result from Rio was the Darwin Initiative whereby Third World taxonomists could receive training from "western" experts; unfortunately, this did not guarantee the replacement of these experts in their turn. Henry Disney himself is an interesting example of achieving much on a shoestring. He has never received a research grant from one of the Research Councils, but in spite of that he has published from his Cambridge base some three hundred papers on scuttle flies (Phoridae), a group as diverse as they are little studied. During the 1990s he made himself unpopular in establishment circles by lobbying successive governments for information about what they were doing for taxonomy beyond rhetoric—and concluding that it wasn't much. Since then a few hopeful signs have appeared. An American initiative opened up the possibility of funding for taxonomy from the National Science Foundation, the biggest teat from which scientists suckle on the west side of the Atlantic. The relevant acronym is PEET—Partnerships for

Enhancing Expertise on Taxonomy—and the result has been that fresh-faced acolytes have been sitting at the feet of grizzled old experts and absorbing their lifetime's experience.

Notwithstanding this hopeful harbinger, the progress towards an inventory of life's richness still seemed too slow. By 2005 it had been estimated that some 1.7 million species had been described—far less than half the total on even the most conservative view. If taxonomists continued to name something like ten thousand new species a year, it would take several centuries to complete the task, and who could tell what might become extinct in the interim? And the latter category might include the taxonomists, too. When you learn that Henry Disney has named about three hundred species of flies it sounds rather a lot, but it is nothing compared with the task ahead for the scuttle flies alone. And most biologists would not want to stop with the name—they would want to go on to the interesting stuff about lifestyles and biology. One particular little fly studied by Henry lives by stealing pollen from the stores of bees—it just requires David Attenborough's photographer to spend a month or two staking out the diminutive creature and you just *know* that fascinating film footage would follow. This is where nature gets really interesting—the name is but a necessary start. But there's no time to linger on just one species—too much to do!

This is where the DNA Barcode comes to the rescue, at least in theory. The laborious business of characterizing a species can be speeded up immensely by using evidence from the genome: effectively DNA "fingerprinting" the species by using a sequence from a diagnostic gene. Formal names are not necessary. Now that determining sequences is fully automated, the whole process of identification can be done with the flick of a phial and the click of a switch. Farewell dusty old Sepia Man with his microscope and library; welcome Barcode Man with his primers and his white coat. Furthermore, since sequences can readily be posted on the internet, the diagnostic sequence can be instantly available around the world. The Genbank site is already an incomparable resource for those many scientists interested in the gene sequences of organisms and what they can tell us about evolution. When a gene is sequenced it is posted on Genbank—it has become a necessary part of the publication process at the time of writing. There are various computer programs which can match a sequence under study to those posted on Genbank, rather like a computer-dating system that looks for a close (though rarely perfect) match. The resource is utilized continuously for everything from researching the latest mutations of viruses to studying the evolution of earthworms. The Natural History Museum aims to "barcode" the British fauna and flora as part of this international endeavour.

There is no question that the gene option has helped to revitalize taxonomy. It would be tragic if it also had the undesirable effect of killing off experts on the "whole organism," by what is euphemistically known as collateral damage. I have watched the number of paid professionals on my own trilobite corner of natural history go down to a quarter in Britain during my working life, and no sign of new jobs for the younger generation. My last three research students have gone on to earn a decent wage in the City. Trilobites are not exceptional. For most of my life I have been an amateur mycologist, and I have watched the few paid experts who can identify mushrooms in fields and woods get older, retire and not be replaced. This is despite the fact that there are still many new fungi to discover. The same is true of professionals who know about less fashionable kinds of insects. There are just so few grants to help the process along. The one beneficial effect of this apparently inexorable change in scientific demographic is that people who know their organisms can continue working usefully into old age. I met the aphid expert Victor Eastop shortly after joining the Natural History Museum; now in his eighties, he is still globe-trotting to identify species of these tiny and troublesome plant suckers. He is a small and neatly bearded man with a prominent nose—quite suitable for smelling out the smallest differences between species. He tells me he has trouble remembering people's names these days, but is still on top of the greenflies. In 2007 the great biologist E. O. Wilson inaugurated a project to collate systematic efforts over the internet, under the title *Encyclopedia of Life.* The project overview states: "Our goal is to create a constantly evolving encyclopedia that lives on the the internet, with contributions from scientists and amateurs alike. To transform the science of biology, and inspire a new generation of scientists, by aggregating all known data about every species." It is an ambitious idea, and one that fits in well with the suggestions made in this book. If it works, maybe we will be just in time to know what there is in the world, before global climate change and population growth put a majority of species under threat. This project is almost as urgent as it is necessary. Museums will be an important part of it.

In the latter part of the eighteenth century, when the Natural History Museum was a part of the British Museum at Bloomsbury, the head of the whole organization was called the Principal Librarian; even the Keeper of Zoology started with the title of Under-Librarian. The collection of books has always been central to the systematic enterprise. The number of books and journals collected by the Museum has grown inordinately: five thousand volumes left the care of Bloomsbury for South Kensington in 1881; there were 750,000 volumes by 1980; they may have doubled again since then. I have explained how the traditional way of determining a species entails comparing a study specimen with a mass of published illustrations and descriptions: the piles of books and papers soon become impressive, and tidiness becomes impossible. The libraries in the Museum are thus working collections as much as those held in the endless ranks of drawers behind the scenes. The libraries are also a repository for treasures. The originals of famous wildlife illustrations are held there, so it is partly an art gallery; rare books are hidden away in secure storage, so another part of the library is a bibliophile's Mecca. The rarest works of natural history are now hugely valuable. That incomparably perfect ornithological masterpiece, John James Audubon's *Birds of America,* published between 1826 and 1838, seldom comes on the market, but when it does it commands vast prices—a set sold for $7 milllion. The Museum once threatened to sell a copy when it experienced a financial crisis. It is surprising how threats of selling off a masterpiece get more publicity than the plight of the beetle collection.

I have mentioned a few of the *florilegia* that combine great artistry with botanical accuracy. The Museum is richly endowed with the drawings of the Bauer brothers, Franz Andreas and Ferdinand Lucas. One could argue indefinitely over which of the two produced the finer illustrations of flowers, or whether both should doff their hats to Georg Dionysius Ehret, who painted the plants that Sir Joseph Banks brought back from Newfoundland and Labrador in the eighteenth century. Despite, or perhaps because of, the perfection of these artists, I find the work of the artist on Captain Cook's first voyage, Sydney Parkinson, particularly affecting. Poor Parkinson died in 1771 before he was thirty, towards the end of the voyage, but not before completing 280 plant paintings and about 900 sketches and drawings. The pencil sketches he made—for example, one of the earliest portraits of the kangaroo—have a delicacy about them that lifts them above mere illustration. The artistic worth of the Natural History Museum's holdings of Sydney Parkinson is perhaps only now being appreciated. Even in the Palaeontology Library the Rare Books Room is lined with brown leather-bound tomes of every shape and size. Here can be found the famous book by George Scrope describing the volcanoes of the Auvergne, alongside a hundred other volumes that have now been largely forgotten. Nor are the books merely of antiquarian interest: an academic visitor from the Czech Republic found a trilobite specimen buried in the collections which turned out to be one of the first ever illustrated from the ancient rocks of Bohemia—and we were able to check the specimen against the figure because a rare and early work by Born was carefully preserved in the library. The rule *never throw anything away* was vindicated yet again. So the library has a collection that includes items more like those in an art or an antiquarian gallery: intrinsically valuable, they are part of a national heritage regardless of their scientific importance. There is an obligation to care for these books and illustrations in perpetuity, so the future of this part of the collection is not in doubt. Treasures need guardians.

The rest of the library comprises more recent books and long runs of journals. The *Philosophical Transactions of the Royal Society* is the Methuselah of them all, having been published since 1665. More and more journals have joined it over the years as the pressure to "publish or perish" has built up, and companies like Elsevier have surfed a wave of academic anxiety to a landfall of profits. It has now become extremely difficult to be up to date, as dozens of ever more specialized journals compete for one's attention. Since the pressure to be first in the field is also increasing, it seems probable that more and more publication will happen initially online. Paper copies can follow on at a leisurely pace, because priority will have been established. Already, most authors send their papers as computer files to their colleagues, who may print them out only if they wish to. It is a profound change in culture. Will libraries continue to be the permanent archive in a virtual world? Will journals in the old sense survive at all? Even the venerable Royal Society has "fast track" journals that appear first online, and then eventually in print. In theory, the contents of an entire library should be accessible over the web—even down to old Scrope himself. It just requires time and money to scan the classic works. I suppose that my attachment to the look of the printed page is hopelessly old-fashioned, as is feeling a curious connection with those who went before me when I take an old volume off the shelf. It might well happen that visitors to the lines of shelves with their runs of journals bound in matching livery will get fewer and fewer as internet access improves. If this happens one might wonder whether the next generation might lose contact with history itself, riding always on a few months' virtual journals, a gathering amnesia erasing the past as intellectual obsolescence creeps inexorably towards what was published the day before yesterday.

One wants to be as positive as one can be about the obvious and accumulating changes in the culture of natural history in national museums. If the difficulties with finding permanent employment for alpha taxonomists are combined with the revolution in library and information access made possible by the web, one comes up with an oddly satisfactory solution. The day of the botanizing vicar will return. The amateur will enjoy a renaissance. In the eighteenth and nineteenth centuries many of the most distinguished names were amateurs, only in the sense that they did not earn a living from working on fungi or fossils. Parsons and moneyed bourgeoisie—Gilbert White or Charles Darwin, say—were typical of this caste. At that time it was possible to hold the relevant literature on a particular group of organisms within a typical middle-class study. Much later, my inspiration, Mr. Morley Jones, still had books on diatoms in his own collection sufficient for him to make a worthwhile contribution to the subject. Today, armed with the unlimited resources of the internet, anyone talented and determined

"Kangaru." Probably the first European drawing of a kangaroo, by Sydney Parkinson, made during Captain Cook's first voyage (1768–71)

enough may carve out a place for himself as an expert on a favoured group of organisms. All the literature and photographs, keys and microscopic details can be made freely available. If the early phase of systematic learning was mostly powered by privilege, the middle phase by support from government for professionals, maybe the third phase will be immensely democratic, and driven by the freedom of information exchange thrown up by the web.

It is already happening: I have met dedicated people in the mycological world who have devoted years to "their" genus of mushroom. I have met devoted and unsalaried bee men and beetle women. The role of the museum will continue to be to house material, and particularly types, but this material will be derived from a different and more inclusive demographic. While the paid professionals move further into molecular work, where the amateur cannot follow, the "hairs on legs" of little-known species will increasingly be studied by a new breed of web-savvy naturalists in contact with all their fellow enthusiasts around the world. Authority will be devolved to a thousand computer terminals. The reign of the solitary and mighty figure below stairs in the Museum may well be coming to an end in the age of *The Encylopedia of Life*. But it will be more important than ever that the professional expert should still retain his or her place as guardian of the collections and as "quality control" on the taxonomy produced by the wider community. There is a division in taxonomy that is nearly as old as the subject itself between "lumpers" and "splitters." The former take a wide view of a species, while the latter tend to recognize more and finely differentiated entities, which are then given separate names.* With regard to birds this may not seem too much of a problem—though there are fierce controversies even here—but with organisms like fungi and molluscs there are unending debates about what makes a species. In actively evolving plants that can also hybridize, the problems are compounded; people have gone prematurely grey trying to wrestle with the complexities of the brambles (*Rubus*). Since evolution is still happening all around us, it would actually be rather surprising if there were *not* such difficulties of definition in nature.

To some, it might seem rather tempting to propose a new species in order to achieve a certain kind of immortality—but it could well be a species that will not stand up to subsequent scrutiny. Many pretty colour forms of mushrooms and shells have been shown to be variants rather than real species. There *has* to be a sound system of refereeing lest "new" species pop up indiscriminately and lead eventually to a riot of unnecessary names. The continued existence of the expert will be absolutely essential to make sure that Linnaeus' original intention of making a system of nature does not founder in a welter of nomenclatural chaos, a return to the Dark Ages of disorder. The professional systematist will be there to adjudicate between new methods of identification carried out by his fellow scientists and data gathered by a grow-

*Molecular evidence can now arbitrate on these cases, and in my experience often seems to find in favour of fine species divisions. There are also increasing numbers of examples of cryptic species, where the molecules are more different than the morphology might indicate. This is particularly the case where populations have been isolated for very long periods of time.

ing army of unpaid experts. It is profoundly to be hoped that scientists in Third World countries will participate in this taxonomic democracy. There need to be more of them to come to care as much about their own fauna and flora as has become commonplace in western countries; if so, there may yet be hope for the future of global biodiversity. The end of all this is to open humankind's eyes and hearts to the joy of our world's biological richness, from microbe to mastodon, from *Selaginella* to *Sequoia*. The means is continued discovery, unravelling the Tree of Life, and describing the fascinating biographies of a million organisms. Every species on Earth has its story to tell. But the first stage will always be the naming of names.

It should be clear from what I have written that I have affection for those people who have worked away unseen behind the public galleries: the secret museum. Without being too fanciful, I might say I have lined them up in my own gallery, curated them by displaying their peculiarities and made of them my own idiosyncratic museum. I could write another book as long as this with a completely different cast that is every bit as extraordinary. A number of my colleagues winked at me when they heard what I was about and hinted that this might be a time to settle old scores. I soon discovered that the real business was to explain what museum science is about, and to try to understand how the taxonomic sciences have evolved since the early days. So I have described a selection of the research that is going on right now. It does not pretend to be a comprehensive account—it's not even representative. It is just my own collection—projects that caught my eye, or seemed to show where science might go, or were chosen just because I admire the people doing them. It is my own Dry Storeroom No. 1.

There is a purposefulness about the scientific benches in the early twenty-first century that is probably more focussed than at any time in the past. After all, we live in competitive times. Formerly, there was more leisure for people behind the scenes to cultivate their eccentricities like prize vegetable marrows, mulching them regularly with their prejudices and fertilizing them with long draughts of solitude. This had something to do with the security of tenure of the established civil servant. I doubt whether a Kirkpatrick, creator of the nummulosphere, or a W. N. P. Barbellion could survive in the present climate.

What I have written about the Natural History Museum in London applies just as much to the American Museum of Natural History in New York, and to the National Museum of Natural History in the Smithsonian Institution in Washington, and to any one of forty similar institutions around the world. All these museums face similar challenges to those I have described to keep systematic research advancing in a changing funding culture. Their research has never been more important at a time when human impact on the environment is causing whole ecosystems to degrade. The great museums may harbour the conscience for the natural world, not merely provide its catalogue. They may be the only places where future generations may be able to find the answer to the question: *What have we done?* The collections may yet shame us all: let us hope history proves otherwise.

For now, what unites the public galleries and the secret hinterland I have explored with you is the celebration of life's richness. We—visitors and researchers alike—are all united with the animals and plants on display, or in their hidden drawers, in the brotherhood of DNA. The fossils are part of our story, be they ever so strange. The least microbe or the greatest mammal deserves our attention. The geological history of our planet is as intimately entwined with the history of the life on its surface as the symbiosis between tree and truffle. Evolution is not a late ingredient to be added as a kind of seasoning to sharpen our taste for the natural world, it is the main course itself. When we wander with the crowds through gallery after gallery of specimens, or pause to examine moving models and videos explaining the working of the eye or the cell division of a bacterium, we are seeing what evolutionary change has accomplished on our planet through nearly four billion years. Our privilege—our uniquely human privilege—is that we alone among the multitude of species can understand the processes that got us where we are today. Those who work in the secret museum on understanding animals and plants remind us, paradoxically, what it is to be human.

I will finish by quoting the diarist W. N. P. Barbellion, even if his evolutionary scenario is no longer quite the ticket. On 22 July 1910 he wrote:

I take a jealous pride in my Simian ancestry. I like to think I was once a magnificent hairy fellow living in the trees and my frame has come down through geological time *via* a sea jelly & worms & Amphioxus, Fish, Dinosaurs & Apes. Who would exchange these for the pallid couple in the Garden of Eden?

Acknowledgements

I am indebted to many friends and colleagues in the Natural History Museum for sharing their oral histories with me, often in return for no more than a good lunch. Any errors that appear in this book are entirely the fault of the author, who may have not been the best of scribes, and nothing to do with those at lunch. I apologize to those who gave me stories that I failed to use: there was no shortage of material. I apologize even more sincerely to those who still have not received their lunch. I should particularly mention the following (in no particular order): Edmund Launert, Sandra Knapp, Peter Hammond, Dick Vane-Wright, Victor Eastop, Martin Hall, Steve Brooks, Andrew Polaszek, Paul Eggleton, David Reid, John Taylor, David Johnson, Vaughan Southgate, Amoret Whitaker, Jenny Bryant, Linda Irvine, Bob Press, Robert Symes, Richard Herrington, Alan Hart, Paul Taylor, Hugh Owen, Ellis Owen, Angela Milner, Cyril Walker, Ollie Crimmen, Ron Croucher, Chris Stringer, Andrew Currant, Alex Ball, Sara Russell, Frances Wall, Chris Stanley, Bob Hutchison, Rowland Whitehead, Klaus Sattler, Kathy Way and Lorraine Cornish. Polly Tucker helped greatly with making Museum archives available to me. Thanks also to Katie Anderson in the Natural History Museum Picture Library. Heather Godwin gave her usual eagled-eyed attention to the first draft, and removed several bad jokes. Robin Cocks read through the manuscript. Jackie Fortey gave indispensable help with picture research.

Further Reading

Arnold, E. N. 2003. *Reptiles and Amphibians of Europe.* Princeton University Press.

Barbellion, W. N. P. 1919. *The Journal of a Disappointed Man.* Sutton Publishing.

Bowler, P. J. 1983. *The Eclipse of Darwinism.* Johns Hopkins University Press.

Cheesman, E. 1949. *Six-Legged Snakes in New Guinea: A Collecting Expedition to Two Unexplored Islands.* Harrap.

Clutton-Brock, J. 1999. *Natural History of Domesticated Animals.* 2nd ed. Cambridge University Press.

Corfield, R. 2003. *The Silent Landscape: In the Wake of HMS* Challenger, *1872–1876.* John Murray.

Erzinclioglu, Z. 2000. *Maggots, Murder and Men.* Harley Books.

Ferguson, N. 2001. *The House of Rothschild: Money's Prophets, 1798–1848.* Penguin.

Gee, Henry. 1996. *Before the Backbone.* Chapman & Hall.

Gilbert, O. L. 2000. *Lichens.* Collins New Naturalist Series.

Godfray, H. C. J., and S. Knapp. 2004. *Taxonomy for the 21st Century.* Philosophical Transactions of the Royal Society B, 359.

Grady, M. 2000. *Catalogue of Meteorites.* Cambridge University Press.

Kenrick, P., and P. R. Crane. 1997. "The Origin and Early Evolution of Plants on Land." *Nature* 389:33–39.

Lawrence, P. N. 1989. *Impressive Depressives: 75 Historical Cases of Manic Depression from Seven Countries.* P. N. Lawrence.

Mayor, A. 2000. *The First Fossil Hunters: Paleontology in Greek and Roman Times.* Princeton University Press.

Molleson, T., and M. Cox. 1993. *The Spitalfields Project. Volume 2: The Anthropology—The Middling Sort.* Council for British Archaeology Research Report No. 86.

Morton, V. 1987. *Oxford Rebels: The Life and Friends of Nevil Story Maskelyne.* Sutton Press.

Ramsbottom, J. 1953. *Mushrooms and Toadstools.* Collins New Naturalist Series.

Rasmussen, Pamela C., and Robert P. Prys-Jones. 2003. *History* vs *Mystery: The Reliability of Museum Specimen Data.* British Ornithologists' Club.

Reid, D. G. 1996. *Systematics and Evolution of Littorina.* The Ray Society.

Rudwick, M. 2006. *Bursting the Limits of Time.* University of Chicago Press.
Russell, Miles. 2003. *Piltdown Man: The Secret Life of Charles Dawson.* Tempus.
Schindler, K. 2005. *Discovering Dorothea.* HarperCollins.
Smith, K. G. V. 1986. *A Manual of Forensic Entomology.* British Museum (Natural History).
Smith, V. S. 2005. *DNA Barcoding: Perspective from a "Partnerships for Enhancing Systematic Expertise (PEET)" Debate.* Systematic Biology.
Spooner, B., and P. J. Roberts. 2003. *Fungi.* Collins New Naturalist Series.
Stearn, William T. 1981. *The Natural History Museum at South Kensington.* Heinemann.
Weil, S. E. 1995. *A Cabinet of Curiosities: Inquiries into Museums and Their Prospects.* Smithsonian Institution.
Whitehead, P. J. P. 1985. *FAO Species Catalogue.* Volume 7. Clupeid Fishes of the World (Suborder Clupeoidei). *An Annotated and Illustrated Catalogue of the Herrings, Sardines, Pilchards, Sprats, Shads, Anchovies and Wolf-herrings.* Part 1—Chirocentridae, Clupeidae and Pristigasteridae. Food and Agriculture Organization, Rome, Fisheries Synopsis, 125, volume 7, part 1:x + 1–304.
Wyse Jackson, P. N., and M. E. Spencer Jones. 2002. *Annals of Bryozoology: Aspects of the History of Research on Bryozoans.* International Bryozoology Association, Dublin.

Illustration Credits

The author and publishers gratefully acknowledge the following sources for permission to reproduce illustrations:

INTEGRATED ILLUSTRATIONS

6 *Diplodocus carnegii. Photo © Natural History Museum, London.*
10 Plesiosaur drawn by Mary Anning, 1824. *Photo © Natural History Museum, London.*
15 Museum cabinets. *Photo © Natural History Museum, London.*
20 Stuffed giraffe specimens. *Photo © Natural History Museum, London.*
21 Tray of molluscs from the Sloane collection. *Photo © Natural History Museum, London.*
34 Fungus gnat *Mycetophila. Photo © Andrew Darrington/Alamy.*
39 Richard Owen with moa skeleton. From Richard Owen, *Memoirs on the Extinct Wingless Birds of New Zealand,* Vol. 2 (London: John Van Voorst, 1879), plate XCVII.
41 Seated statue of Charles Darwin. *Photo © Natural History Museum, London.*
45 Title page of *Stray Feathers, A Journal of Ornithology for India and Its Dependencies,* Allan Octavian Hume, Vol. 2 (1874).
54 Fossil ostracode *Colymbosathon ecplecticos. Photo © David Siveter. Article ©* The Sun.
57 Orobatid mite larva *Archegozetes. Photos © Richard Thomas.*
62 Duck-billed platypus. *Photo © Natural History Museum, London.*
64 Nematode worm *Coenorhabditis elegans. Photo © Phototake Inc./Alamy.*
65 Edible black truffle. *Photo © Jackie Fortey.*
76 Trilobite "mines" in Morocco. *Photo © Brian Chatterton.*
78 Devonian trilobite *Erbenochile. Author's own collection.*
79 Spiny trilobite, odontopleurid. *Photo © Brian Chatterton.*
89 Ammonite. *Photo courtesy M. K. Howarth.*
95 Colin Patterson. *Photo courtesy Peter Forey.*

98 Kenneth Oakley, 1953. *Photo by Daniel Farson/Picture Post © Getty Images.*
101 *The Examination of the Piltdown Skull*, by John Cooke. *Photo © Geological Society, The/NHMPL.*
102 Piltdown artefacts. Drawings from the *Quarterly Journal of the Geological Society of London*, 1914.
109 Palaeontology laboratory. *Photo © Natural History Museum, London.*
112 Lycopsids from South Island, New Zealand. *Author's own collection.*
117 Winkles *Afrolittorina, Austrolittorina. Photo courtesy David Reid.*
119 Lucinid clam *Rasta thiophilia. Photo courtesy John Taylor.*
120 Clam *Plicolucina flabellata. Photo courtesy John Taylor.*
121 View of the Chelsea Physic Garden c. 1910. *Photo © RBKC, Libraries.*
126 John Peake. *Drawing courtesy John Taylor.*
128 Nineteenth-century print of Loch Ness. *Author's own collection.*
130 Peter Whitehead. *Photo courtesy Oliver Crimmen.*
132 Peter Whitehead. *Article* © The Sunday Times.
140 Nummulosphere. Illustration from R. Kirkpatrick, *The Nummulosphere*, London, 1913.
149 Peter Purves and whale carcass. *Photo © Natural History Museum, London.*
152 Museum office. *Photo © Natural History Museum, London.*
155 Herbarium. *Photo © Natural History Museum, London.*
156 Traveller's joy from Leonhard Fuchs, *De Historia Stirpium*, Basel, 1542.
170 Larkspur. *By permission of the Linnean Society of London.*
179 Diatoms from Christian Ehrenberg, *Mikrogeologie*, 1854.
181 Lake Baikal. *Photo © David Williams.*
186 Screw worm (*Cochliomyia hominivorax*). *Photo courtesy Martin Hall.*
188 Head of the larval screw worm (*Cochliomyia*). *Photo courtesy Martin Hall.*
191 Screw worm plaque. *Photo courtesy Martin Hall.*
193 Bruce Frederic Cummings. *Photo courtesy Eric Bond Hutton.*
196 Walter Rothschild. *Photo © Natural History Museum, London.*
198 *Butterfly collector* by Ian Jackson, *Punch*, 22 May 1985. *Reproduced with permission of Punch Ltd.*
201 Dick Vane-Wright. *Photo courtesy Dick Vane-Wright.*
204 (*Top*) Decomposing piglet. (*Bottom*) Bluebottle (*Calliphora vicina*). *Photos courtesy Amoret Whitaker.*
209 Springtail. *Photo © Holt Studios International/Alamy.*
213 Head of midge *Heterotrissocladius grimshawi. Photo courtesy Steve Brooks.*
214 (*Top*) Beetle display. *Photo © Natural History Museum, London.* (*Bottom*) Rhinoceros beetle. *Photo courtesy Rob Knell.*
217 Field laboratory in Africa. *Photo © Dick Vane-Wright.*
218 Beetle *Agathidium vaderi. Drawing courtesy Quentin Wheeler.*
224 (*Top*) Martian meteorite ALH 84001. *Photo courtesy NASA/JSC.* (*Bottom*) Micrograph of ALH 84001. *Photo courtesy NASA.*

231 Jadarite. *Photo courtesy Chris Stanley.*

234 (*Top*) SEM image of Cenozoic bryozoan (*Exochella jellyae* Brown, 1952). (*Bottom*) Drawing by Brown of the holotoype. *Photos courtesy Paul Taylor.*

243 Mineral Gallery, Natural History Museum. *Photo © Natural History Museum, London.*

245 Portrait of Edward Heron-Allen. *Photo courtesy Clive Jones.*

248 Big Hole diamond mine, Kimberley, South Africa. *Photo © Chris Howes/Wild Places Photography/Alamy.*

250 Diamond-cutting machine. *Photo © Corbis.*

253 Native silver. *Photo © Natural History Museum, London.*

254 Arthur Russell. Photo from *Nature Stored, Nature Studied,* British Museum (Natural History), 1981.

261 Cartoon of Edwin Ray Lankester riding an okapi. *Punch,* 12 November 1902.

263 Sir Gavin de Beer. *Photo © Natural History Museum, London.*

267 Museum logo. *Author's own collection.*

271 Extinct mammal *Myotragus. Photo © Natural History Museum, London.*

273 Lucy Evelyn Cheesman. *Photo © Natural History Museum, London.*

277 Model of the dodo. *Photo © Natural History Museum, London.*

279 Meinertzhagen collection. *Photo © Natural History Museum, London.*

282 Museum storage in Surrey caves, 1943. *Photo by Fox Photos/Hulton Archive © Getty Images.*

283 War damage to the Botany Department. *Photo © Natural History Museum, London.*

286 House journal *Tin Hat. Author's own collection.*

288 HMS *Challenger. Author's own collection.*

296 Richard Owen carving a dodo. *Photo © Science Museum/Science & Society Picture Library.*

297 W. D. Lang. *Photo courtesy The Royal Society.*

298 Cretaceous bryozoan *Pelmatopora. Photo courtesy Paul Taylor.*

301 Carpoid *Cothurnocystis elizae* Bather. *Photo © Natural History Museum, London.*

310 Drawing of a kangaroo by Sydney Parkinson. *Photo © Natural History Museum, London.*

PLATES

I

1 Front façade of the Natural History Museum. *Photo © Natural History Museum, London.*

2 Main hall, Natural History Museum. *Photo © David Pearson/Alamy.*

3 *Sir Richard Owen* by William Holman Hunt, 1881. *Photo © Natural History Museum, London.*

4 Ceiling panel, detail of *Pinus sylvestris. Photo © Natural History Museum, London.*

5 Giant sequoia. *Photo © Natural History Museum, London.*

6 Moonfish in jar. *Photo © Natural History Museum, London/Alamy.*

7 Osteology storeroom. *Photo © Natural History Museum, London.*

8 Nathan Muchhala, Bellavista Cloud Forest Reserve, Ecuador. *Photo © Jackie Fortey.*

9 Nectar bat (*Anoura fistulata*). *Photo © Nathan Muchhala.*

10 Collecting insects. *Photo © Natural History Museum, London.*

11 The Peacock Fountain, Christchurch Botanic Gardens, New Zealand. *Photo © Jackie Fortey.*

12 Portrait of Linnaeus by Martin Hoffman. *By permission of the Linnean Society of London.*

13 "Old Man Banksia." *Photo © Natural History Museum, London.*

14 Natural History Museum library. *Photo © Natural History Museum, London.*

15 The DNA laboratory. *Photo © Natural History Museum, London.*

16 Living bryozoan (*Adeona*). *Photo courtesy Piotr Kuklinski.*

17 Ostracod. *Photo courtesy David Siveter.*

18 Moroccan trilobites for sale. *Photo © Brian Chatterton.*

19 Reconstruction of *Archaeopteryx. Photo © Natural History Museum, London.*

20 *Archaeopteryx. Photo courtesy Angela Milner © Natural History Museum, London.*

21 Boxgrove excavation. *Photo courtesy The Boxgrove Project.*

22 Suffolk cliffs. *Photo © Rob Francis.*

23 Cichlid fish. *Photo © 2007 Johnny Jensen/Image Quest Marine.*

24 Nummulites. *Photo © Dr. Basil Booth/GeoScience Features Picture Library.*

25 Sea spiders. *Photo courtesy Derek Siveter.*

II

1 Portrait of Sir Joseph Banks by Thomas Phillips. *By permission of the Linnean Society of London.*

2 Herbarium specimen of the cocoa plant. *Photo © Natural History Museum, London.*

3 Lichens on gravestone. *Photo © Jackie Fortey.*

4 Herbarium specimen of sweet pea. *By permission of the Linnean Society of London.*

5 Tomato relative *Solanum huaylasense* Peralta. *Photo © Blanca Léon.*

6 Diatoms. *Photo Dr. Neil Sullivan, University of California © NOAA.*

7 *Hypericum* species. *Photo © Jackie Fortey.*

8 Dr. Miriam Rothschild. *Photo © Tony Evans/naturepl.com.*
9 Insect pest *Aleurocanthus woglumi. Photo © Gillian Watson.*
10 Parasitic wasp *Encarsia perplexa. Photo © Andrew Polaszek.*
11 Field laboratory. *Photo © Dick Vane-Wright.*
12 Termite mound, Kakadu National Park, Australia. *Photo © Rob Francis.*
13 Termite *Bifidtermes. Photo courtesy Paul Eggleton.*
14 Moth and butterfly collection of Sir Alfred Russel Wallace. *Photo © Natural History Museum, London.*
15 Thin section of olivine basalt. *Photo M. Hobbs/GeoScience Features Picture Library.*
16 Deep-sea tubeworms. *Photo © NOAA.*
17 Deep-sea sulphide chimney. *Photo © NOAA.*
18 Oldoinyo Lengai volcano. *Photo © Rob Francis.*
19 Nuratau Mountains, Uzbekistan. *Photo © Jan Sevcik.*
20 Kovdorskite. *Photo courtesy Frances Wall.*
21 Koh-i-noor diamond mould. *Photo courtesy Alan Hart.*
22 Bournonite. *Photo © Natural History Museum, London.*
23 Koh-i-noor diamond. *Photo © The Royal Collection © 2007 Her Majesty Queen Elizabeth II.*
24 Latrobe gold nugget. *Photo © Natural History Museum, London.*
25 The "accursed amethyst." *Photo courtesy Alan Hart.*
26 Illustration of owls, Plate 432 from John James Audubon's *Birds of America* (1835–38). *Photo © Natural History Museum, London/Alamy.*
27 Gymea lily from *Illustrationes Florae Novae Hollandiae,* 1816, by Ferdinand Bauer. *Photo © Natural History Museum, London.*
28 Seaside ragwort from Newfoundland, Volumes, sketches by Georg Dionysius Ehret. *Photo © Natural History Museum, London.*

While every effort has been made to trace the owners of copyright material reproduced herein, the publishers would like to apologize for any omissions and would be pleased to incorporate missing acknowledgements in any future editions.

Index

Page numbers in *italic* refer to illustrations and captions.

Acharius, Erik (1757–1819), 161
Afghan snowfinch (*Montifringilla theresae*), 279
African giant hog (*Hylocheorus meinertzhageni*), 279
Agathidium genus (slime beetles), 217–18
Agricola, Georgius, 302; *De rerum metallica,* 229
AHOB (ancient human occupation of Britain), 104–6
Akihito, Emperor of Japan, 49
Albert, Prince Consort, 38, 42, 250, 252
Aleurocanthus woglumi (white fly species), 198
Alexander (American entomologist), 216
algae, and lichens, 159, 163–4
Alvin (deep submergence vessel), 237
American Museum of Natural History, New York, 279, 313
amethyst, 245–7
ammonites, 85, 87–8, 89–90, *89,* 299*n*
Anagyrus lopezi (parasitic encyrtid wasp), 199
Anning, Mary, 10, *10*
ants, 206–7
aphids, 197
Archaeopteryx lithographica (Jurassic bird), 14, 81–4, 94–5, 107, 263
Archegozetes larva, *57*
Arkell, W. J., 89, 94
Arnold, Nick, 151
Arthropoda, 17, 150, 274
ascomycetes, 163
Ashton Wold (estate), near Peterborough, 195
Atropa belladonna (deadly nightshade), 172
Audubon, John James: *Birds of America,* 307
authority, individual, 35–6
automated sequencer (machine), 141

Bacchus, Jim, 215
Bacon, Francis, 139
bacteria, 35, 237
Baikal, Lake (Russia), 181–2, *181*
Bairstow, Leslie (1907–95), 85–8, 93, 124
Ball, Alex, 233
Ball, Harold William ("Bill"), 17, 96
Banks, Sir Joseph: collection, 16; plants from Newfoundland and Labrador, 308; on plants of utility and virtue, 43
Barbellion, W. N. P., *see* Cummings, Bruce Frederic
barnacles, *see* cirripedes
Baryonyx walkeri (dinosaur), 107–10, *109*
Bate, Dorothea (1878–1951), 270–2, *271,* 293
Bather, F. A., 299
bats, 33–4
Baudrillard, Jean, 294
Bauer, Ferdinand Lucas (1760–1826), 307
Bauer, Franz Andreas (1758–1840), 307
Beagle, The (Darwin's ship), 50, 179, 259, 287
beetles: characteristics, 215, 217–18; number of species, 34, 213, *214,* 216
Belcher, Phoebe, 17–18
belemnites, 85
bettongs, 68
Beuys, Joseph, 157
Bilharz, Theodore Maximilian (1825–62), 143
bilharzia, 143–5
biodiversity, 32, 48, 81*n,* 286–9, 303, 312
birds: and *Archaeopteryx,* 83; research downgraded at Museum, 125

Bishop, Clive, 266
black smokers, 237–8
Blake, William, 197
Bletchley Park, 166
blowflies, 201–2
bluebottles (*Calliphora vicina*), 202, 203
blue whale (*Balaenoptera musculus*), 147
Blyth's kingfisher (*Alcedo hercules*), 279
bodies, *see* corpse decay
Bodmer, Sir Walter, 124, 138
Boehm, T. E., 259
Bolton, Barry, 200, 206–7, 210; *Bolton's Catalogue of Ants of the World,* 207; *A New General Catalogue of the Ants of the World,* 207
Borneo, new species discovered, 47
botany, and nomenclature, 166–8
Botany Department, 24–6, 159, *283*
Boulenger, Georges Albert (1858–1937): *Catalogue of the Freshwater Fishes of Africa in the British Museum,* 134
bournonite, 255
Boxgrove, Sussex, 104–5
Boxshall, Geoffrey, 150
Bragg, Sir Lawrence, 228
brambles (*Rubus*), 311
Bridges, J., 225–6
Britain, ancient human occupation, 104–7
British Empire: collectors, 43–4, 46; expansion, 43–4
British Museum Act (1753), 13
Brookes, Steve, 212
Brunfels, Otto (1488–1534), 155, 167
Brunfelsia, 157
Bryant, Jenny, 178
bryozoans, *234,* 296–9
Bryson, Bill: *A Short History of Nearly Everything,* 183
Bulinus (snail group), 145
burying beetle (*Nicrophorus* species), 203
Bush, George W., 218–19
Buskerud, Norway, silver mines, 252
butterfies (*Lepidoptera*), 22, 44; collections and collectors, 205–6

Calcutta, Botanical Garden, 47
Calliphora vicina, see bluebottles
Calliphoridae Family, 202
Cambridge University, science collections, 43
Cameroonian tumbu fly (*Cordylobia anthropophaga*), 187
Candid Friend (journal), 260
carbonatites (volcanic rocks), 240–2
Carlyle, Thomas, 206
Carnegie, Andrew, 5, 7
Carnegie Museum, Pittsburgh, 43
carpoids (fossils), 299, 300, *301*
cassava (*Manihot esculenta*), 199
Central America, 173
Cetacea (mammal group), 146*n*
Challenger, HMS, 287, *288,* 289
Chalmers, Sir Neil, 123–5, 266, *267,* 275
Chambers, Jim, 136
Charig, Alan, 109
Chater, Arthur, 183
Chatterton, Brian, 74
Cheesman, Lucy Evelyn (1881–1969), 272–3, *273,* 287
Chelsea Physic Garden, *121*
Chesters, K. I. M. (*later* Elliott), 110
chironomid midges, 212–13
chocolate, 157
chordates, 299–300
Christchurch, New Zealand, Botanical Garden, 47
Chrysomyia bezziana, 190
cichlid fish, 133–5
cirripedes (barnacles), 90–1
cladistics, 69–71, 94
clams, 117–18, *119, 120*
Claringbull, Sir Frank, 257, 264–5
Clark, Ailsa M., 275
Clay, Theresa, 279–80
Clematis vitalba, 156
climate: change, 112–13; and human habitation, 106
Clinton, Bill, 223
Clutton-Brock, Juliet, 275
Clydoniceras discus (ammonite), 87
Clypeus ploti (sea urchin), 88
Cochliomyia hominivorax, see screw worm
cockroaches, 211
coelurosaurs, 83
Coenorhabditis elegans, see nematode worm
coffin fly (*Conicera tibialis*), 203
Coleman, Arthur P., 8

Colymbosathon ecplecticos (ostracode), 53, *54*
computer methods, 289–90
conmen and fraudsters, 277–81
Conus (snail genus), 70
Convention on Biodiversity (1992), *see* Rio de Janeiro
Cook, Captain James, 43, *310*
Coope, Russell, 106
Cope, Edward Drinker, 80
copepods, 150
Coquerel, Charles, 188
Corbet, Gordon, 199–200
Corfield, Richard: *The Silent Landscape,* 287
corpse decay, 201–5, *204*
Cothurnocystis elizae, 301
Cox, L. R., 108
coypu (*Myocastor coypus*), 100
crane flies (Tipulidae), 216
creationism, 40–2
Creation Museum, Kentucky, 42
Criddle, Alan, 230, 230*n*
Crimmen, Ollie, 137, 257
Croucher, Ron, 107–9
Cryptocercus (wood-eating cockroach), 211
cryptogams (plants), 159, 161, 164–5
cryptozoologists, 126–7
crystallography, 227–8, *231*
Culpeper, Thomas, 153
Cuming, Hugh, 44
Cummings, Bruce Frederic (W. N. P. Barbellion; 1889–1919): on entomologists, 200; on evolution, 313; *The Journal of a Disappointed Man* (by "W. N. P. Barbellion"), 192; life and career, 192–4, *193*, 313
Curds, Colin, 199
Curie, Marie, 272
Currant, Andy, 106

Darwin, Charles: amateur status, 309; barnacle collection, 90–1; collection, 16; on human descent, 97; and Meinertzhagen, 278; *The Origin of Species,* 91; Owen opposes evolutionary theory, 38, 42; presents first draft of Theory of Evolution at Linnean Society, 170; reputation, 152–3, 171; statue, 40, *41*, 259; voyage on *Beagle,* 50, 179, 287; *see also* evolution, theory of
Darwin, Erasmus, 153
Darwin Centre, 29, 123, 135, 140*n,* 150, 152
Darwin Initiative, 181, 304
Dawson, Charles, 97, 99, 102–4
Day, Mick, 216
Dean, Bill, 8
de Beer, Sir Gavin, 37, 98, 123, 262–4, *263,* 266–7
de Beer family, 248–9
decomposition, *see* corpse decay
Destructive Imported Animals Act (1932), 100
diamonds, 227, 241, 247–51, *248*
diatoms, 178–82, *181*
dichotomising keys, 175
Dillenius, J. J.: *Hortus Elthamiensis,* 167
dinosaurs: displayed, 5–7; feathered, 83–4; reconstruction, 108–9, 295
Diplodocus carnegii (dinosaur), 5–7, *6*
Diptera Order (of insects), 185–6
Disney, Henry, 304–5
Dixon, Michael, 266–7
DNA, 62–3, 141–2, 290, 313; Barcode, 305
dodo, 276, *277*
Doyle, Sir Arthur Conan, 99–100, 103
Dracula, *198*
Dry Storeroom No. 1 collection, 29–30
Dublin, Natural History Museum, 293
duck-billed platypus (*Ornithorhynchos*), 61, *62*
Durbidge, Paul, 105
Durrant, John, 282

Eastop, Victor, 306
Eberhardie, Brian, 75, 77
echinoderms, 96, 299–300
ecology, 32
ecosystems, under threat, 47, 303–4
Edward, Prince of Wales (*later* King Edward VII), 259
eels, 126
Eggleton, Paul, 210
Ehrenberg, Christian Gottfried (1795–1876): *Mikrogeologie,* 180, *181*
Ehret, Georg Dionysius (1708–70), 167, 308

Einstein, Albert, 260
Electron Spin Resonance, 106
Eliot, T. S., 165, 197
Elliott, Graham, 110
Encyclopedia of Life, The, 306, 311
Endeavour (Cook's ship), 43
Entomology, Department of, 21–3
environmental changes, 292
Eoanthropus dawsoni, 97
Erbenochile erbeni (trilobite), *78*, 79, 84
Erfoud, Morocco, 221
Erzincliogglu, Zakaria: *Maggots, Murder and Men*, 203
Eucalyptus trees, and false truffles, 68
evolution, theory of: and descent of life, 40–2; disputes over, 40–2, 153; essential nature, 313–14; first draft presented at Linnean Society (1858), 170; Flower promotes, 259; Owen opposes, 38, 259; and stratigraphy, 93
exaptation, 84
Exochella jellyae, *234*

Fagan, Charles, 262
family, in scientific classification, 50
Faraday, Michael, statue, 40
feathers, on dinosaurs, 83–4
Ferguson, Niall, 194
Ferris, Colonel W., 246
First World War (1914–18), 281–3
Fisher, Geoffrey, Archbishop of Canterbury, Baron Fisher of Lambeth, 127
fishes, 129–38
Fitzroy, Captain Robert, 287
fleas, 194–5
Fletcher, Lazarus, 222
flies, *see* Diptera Order
Flinders, Matthew, 232
Flora Mesoamericana, 173–4
Flora Zambesiaca, 174
florilegia, 307
Flower, Sir William (1831–99), 39, 147, 259, 261
flowering plants, evolutionary tree, 110
fluorspars, 254–5
fly agaric, *34*
Foraminifera, 138
forensic entomology, 202–5, *204*
Forensic Science Research Centre, Durham University, 203
Forey, Peter, 96
formalin, destructive effect on DNA, 142
Fortey, Richard, *152*
fossils: appearance, 12–14; Bairstow and, 85–6; continuing discoveries, 112; as evidence of evolution, 93; identification, 35; importance, 313; preparation of, 108; and stratigraphy, 87–8, 90, 93
Fowles, John: *The Collector*, 205
Frankhawthorneite, 230*n*
Fraser, F. C. (1903–78), 148
fraudsters, *see* conmen and fraudsters
Fries, Elias (1794–1878): *Systema Mycologicum* (1821–32), 53
Frozen Ark project (FARC), 143
frozen-tissue collection, 142–3
Fuchs, Leonhard (1501–66), 51, 157*n*; *De historia stirpia*, 155
Fuchsia, 157
fungi, 53, 159–61, 163–5; *see also* truffles
fungus gnats (*Mycetophila*), 33–4, *34*

Galileo Galilei, 153
Gardiner, Brian, 171
Gee, Henry: *Before the Backbone*, 300
Genbank, 305–6
General Library, 27–8
genome, 42, 62–3, 110–11, 142, 151, 173
Gentry, Alan, 124
Geographic Information Systems (GIS) technology, 289
Geological Society of London, 87, *101*, *102*, 169–70
Geological Survey, 252
George VI, King, 264
Gerarde, John: *Herball, or General Historie of Plants*, 157
giraffes, *20*
Gladstone, William Ewart, 38
Godwin-Austen, Henry Haversham, 44
Goldsworthy, Andy, 294
Gould, Stephen Jay, 84; *The Panda's Thumb*, 139
Grady, Monica, 222–3
graphite, 227
gravestones, and lichen analysis, 163

Greenwood, Humphrey, 133–4, 135–6, 139; *The Cichlid Fish of Lake Victoria: Biology of a Species Flock,* 135
Gregory, Richard, 269
Günther, Albert Carl Ludwig Gotthilf, 134

Haldane, John Burdon Sanderson, 33, 262
Hales, Arthur, 285
Hall, Martin, 185–6, 188–9
Hamilton, Sir William, 232
Hammi (Berber of Erfoud, Morocco), 75–6
Hammond, Peter, 215–16, 288
Harding, John, 274
Hardy, Thomas: *Return of the Native,* 221
Harker, John, 233
Harmer, Sir Sidney (1862–1950), 147
Hart, Alan, 244–5, 247, 251, 252
Hartnell, William, 207
Hatchett, Charles, 255
Hawaii, 288
Hawking, Stephen, 260
Hayward, Arthur, 276
head louse (*Pediculus humanus*), 190, 192
Hedley, Ron, 233, 265
hemichordates, 300*n*
Herbarium, 24–5, *155*
Heron-Allen, Edward (1861–1943), 245–7, *245*, 247, 251
herpetologists, 151
Herrington, Richard, 238–9
Heterotrissocladius grimshawi, 213
Hilda, St., Abbess of Whitby, 85
Hinton, Martin (1883–1961), 100–1
Hirst, Damien, 135
Hollis, David, 216
Holmes, Arthur, 240
holotypes, 60, 61, 82
Homo heidelbergensis, 105
Homo sapiens: and climate, 107; evolution of, 104–5; life span as species, 23
Hopwood, Arthur (1897–1969), 101
Horizon (BBC TV programme), 257, 264
Hoskins, George, and Miriam Rothschild: *An Illustrated Catalogue of the Rothschild Collection of Fleas (Siphonaptera) in the British Museum (Natural History),* 195
Howarth, Graham, 210
Hughes, Ted, 183
Humboldt Museum, Berlin, 32
Hume, Allan Octavian: *Stray Feathers,* 44, *45*
Humphries, Chris, 289
Hunt, William Holman, 8
Hunterian Museum, Glasgow, 43
Huperzia (Lycopodiacea genus), 111
Hutchison, Bob, 222, 225–6
Huxley, Thomas Henry, 40, 259
Huys, Rony, 150
Hylocheorus meinertzhageni, see African giant hog
Hypericum (St. John's wort genus), 183
hypothesis-driven research, 298–301, 303

ice age, 106
ichthyosaurs, 9
Indian Geological Survey, 46
Indian Museum, Calcutta, 46–7
insects: collecting, 216, *217*; species variety and numbers, 22–3, 33–4, 206; specimens, 196–7
International Code for Zoological Nomenclature, 80
International Commission on Zoological Nomenclature (ICZN), 81
International Entomological Congress, Brazil (2000), 213
International Mineralogical Association, 229
internet (web), 174–7, 306, 309–11
Island of Mull, The: A Survey of the Environment and Vegetation, 165
isotopes, 106

Jacobs, Jane, 269
Jadarite, 230, *231*
James, Peter, 161
Jarvis, Charlie, 168, 171
Jefferies, Dick, 96, 299–301
Jenner, Edward, 189
Jermy, Clive, 200
Johnson, David, 144–5
Johnston, Sir Harry, 260
Journal of Zoological Nomenclature, 81*n*
Joyce, James, 192

Kaempfer, Engelbert (1651–1716), 158
kangaroo, 308, *310*
Kansas, University of, *see Treatise on Invertebrate Paleontology, The*
Kennedy, W. J., 90
Kenrick, Paul, 110, 303
Kew, Royal Botanic Gardens, 159, 161, 165, 184
Kimberley, South Africa, *248*
Kingsbury, Arthur (1906–68), 280–1
Kirkaldy, G. W., 52
Kirkpatrick, Randolph (1865–1950), 138–9, 313; *The Nummulosphere*, 138–9, *140*
Knapp, Sandra (Sandy), 171–3, 174, 303
Knipling, Edward (1909–2000), 189
Koh-i-noor diamond, 249–51, *250*
Kola Peninsula (Finland/Russia), 241–2
Kongsberg silver wires, 252, *253*
Kryptonite, Green, 230

laboratory methods, 106–7
ladybirds, 215
Lambshead, John, 151
Lamont, Archie, 58
Lang, W. D. (1878–1966), 92–3, 296–9, *297*, *298*
Lankester, Sir Edwin Ray (1847–1929), 99, *101*, 259–62, *261*, 291; *Diversions of a Naturalist*, 262; *Textbook of Zoology*, 260
Lanthanum, 242
larder beetle (*Dermestes*), 203
Larkspur (*Delphinium consolidum*), *170*
Larus ridibundus (black-headed gull), 48
Latin language, in scientific names, 49–51
La Trobe, Charles Joseph, gold nugget, 252
Launert, Edmund, 174, 178, 200, 209, 264
Lawrence, Peter, 208–10
Lawrence, T. E. (Lawrence of Arabia), on Meinertzhagen, 280
L chondrites (meteorites), 226
lead pollution, 161–2
Legg, Julian, 266
Leptodactylus fallax (Caribbean mountain chicken), 143
Leverhulme, William Hesketh Lever, 1st Viscount, 104
libraries, 27–8, 56, 58–9, 307–9
Libya, screw worm found and eliminated, 189
lice, 190, 192
lichenometrics, 163
lichens, 161–5
Linnaean Plant Name Typification Project, 167–8
Linné, Carl von (Linnaeus; 1707–78): classification and naming of plants and animals, 49–52, 70, 166–9, 177, 311; and cryptogams, 159, 166; herbarium, *170*; *Hortus cliffortianus*, 167; and human louse, 190; *Species Plantarum*, 53; Swedish background, 53; *Systema Naturae*, 53
Linnean Society of London, 169–71
Lister, Adrian, 106
Little, Crispin, 238–9
Littlewood, Tim, 142
Livingstone, David, 134
lizards, 151
Loch Ness Monster, 126–8
London Biodiversity Partnership, 136
Lycopodiella (genus), 111
Lycopodium (genus), 111, *112*
lycopsids, 110–11, *112*

Mabey, Richard: *Flora Britannica*, 48
machines and technology, 233–6
Madeira, Flora, 174
maggots, 185–7, 202, 282
Magnitogorsk, Russia, 238–9
Mair Jones, Mrs. (*née* Heron-Allen), 245–6
Major, (Sir) John, 181
Majorca, Cuevas de los Colombs, 270
Malawi, Lake (Africa), 133–4
Malay Peninsula, vulnerability and protection, 289
Mammuthus primigenius (mammoth), 13
man, *see Homo sapiens*
Manton, Irene, 274
Manton, Sidnie (*later* Harding; 1902–79), 273–4; *Arthropoda*, 274
Marine Isotope Stages, 106*n*
Mars: life-forms from, 223–5; meteorites from, 220–1, 223, *224*, 237*n*
Marsh, Othniel Charles, 80
Marx, Karl, funeral, 262
Mass Spectrometric Uranium series, 106
Mattingly, Peter, 207–8
Mayor, Adrienne, 293

Meade, Johnny, 92–3
Meinertzhagen, Colonel Richard (1878–1967), 277–81, *279*, 287
Mellersh, Admiral A., 259
meteorites: effect on trilobites, 226; from Mars (SNC), 220–1, 223, *224*, 236; nickel-iron, 221–2
Meyrick, Edward, butterfly collection, 44
mica, 228
Microlepidoptera, 282
microsatellites, 63
Miles, Roger, 269
Millennium Seed Bank Project, 184
Miller, Kelly, 218
Milner, Angela, 83, 109
Mineralogy, Department of, 28, 222, 226–7, 229, 233, *243*, 244
minerals: naming, 230–1; structure, 228–9; taxonomy, 227–8, 229–30
Missouri Botanic Garden, 173
Mitchell, Edward, 92
mitochondrion, 63–4
Moa (*Dinornis maximus*), 38, *39*
Molecular Biology Laboratory, 140–1
molecular sequencing, 141–2, 217
molecular techniques, 62–5, 164, 290, 311*n*
Molleson, Theya, 104, 275
molluscs, 19–20, *21*, 115–18
Monticchio Lakes, Italy, 241
Monticelli collection (rocks), 232
Montifringilla theresae, see Afghan snowfinch
Moon, rock from, 231
Morley Jones, A., 180, 309
Morley Jones, Catherine, 180
Morris, Harry, 103
Morris, Sam, 18
Morris, William, 157
Morrison-Scott, Sir Terence, 264
Morton, Sir William, Agreement (1961), 165, 174
Morzadec, Pierre, 74, 79
mosquitoes, 207–8
Mound, Lawrence, 266
Muir-Wood, Helen (1895–1968), 91, 93
Mull, Isle of, 165
Murray, John, 287
museums: proliferation and rationale, 42–3; research, 313; role and function, 292–5, 310–11
mushrooms, relationship to truffles, 67–8
muskrat (*Ondatra zibethica*), 100, 262*n*
Mutch, Bob, 105
Mycetophila, see fungus gnats
mycology, *see* fungi
Myotragus (mouse-goat), 270–1, *271*
myxomatosis, 195

Nabokov, Vladimir, 205
Naggs, Fred, 18, 19
names (scientific): allotting and validating, 56, 58–60, 79–81, 312–13; binomial system, 50–1, 166–9; international agreement on, 48–9; Latin and Greek forms, 49–51; and PhyloCode, 70–1; witty and apposite, 52–3
Naratau Mountains, Uzbekistan, 241
Nares, Captain John, 287
NASA (USA), 223
National Science Foundation (USA), 304
Natural Environment Research Council (UK), 302
Natural History Museum: appearance and structure, 4–5; appointment of Director, 123; character, 3; collections moved to present premises, 37–8; danger from pests, 158–9; diary, 17–18; Directors, 258–67; displays, 293–5; disposition of departments, 18–22; Earth Galleries, 294; extensions and modifications, 29; hierarchy and titles, 91–4; *Horizon* TV programme on, 257, 264; Keepers, 17; libraries, 307–9; logo, 267, *267*; management changes, 124–6, 130, 266; mineral gallery, *243*; new dinosaur galleries, 122; offices, *152*; report on administration (1904), 261–2; Research Council status ends, 265–6; research funding, 302–3; Russell Room, 253–6; science grants, 268; security routines, 10–11; shopping and educational areas and activities, 268–9; staff dress code, 37, 92; staff qualifications and recruitment, 7–8; supposed illicit still in model of whale, 275–6; takes over freehold, 268; title and separation from British Museum, 16; total number of specimens, 46; towers, 201; Trustees, 120–4, 126–7, 261–2;

Natural History Museum *(continued)*:
in two world wars, 281–5, *283*; Waterhouse designs, 38, 293; women in, 269–75; working routines, 18
Nature (journal), 83, 128
nematode worm (*Coenorhabditis elegans*), 64, *64*, 151–2, 302
Ness, Loch, *128*
Newman, Barney, 276
Newton, Sir Isaac: reputation, 153; statue, 40
Nile perch (*Lates niloticus*), 133–4
Niobium, 255
nits (head lice egg cases), 192
nummulite (fossil), 138–9, *140*

Oakley, Kenneth P. (1911–87), *98*, 99, 104, 275
ocean floors, and VMS deposits, 237–8
Office of International Epizoology, Paris, 190
Ohlins, Wolf, 267
okapi (*Okapia johnstoni*), 260, *261*, 292
Oldoinyo Lengai (volcano), Africa, 240
Optically Stimulated Luminescence, 106
orchids, 184
Order Out of Chaos: Linnaean Plant Names and Their Types, 168
Orientobilharzia, 145
orobatid mite larva (*Archegozetes*), *57*
orthogenesis, 297
osteology collection, 29
Owen, Ellis, 91, 94, 291
Owen, Sir Richard (1804–92): acquires *Archaeopteryx* for British Museum, 82; and administration, 261; bronze statue, 40; entertaining, *296*; establishes separate museum for Natural History, 37–40, 259; opposes Darwin, 38, 42, 259; pictured with moa skeleton, *39*; portrait, 8–9; and Trustees, 123
Owens, Bob, 300
Oxford University, museum of natural history, 43

Paddock, Barrie, 180
Paine, Tom, 118–19
Pakefield, Suffolk, 105
Palaeontographica Indica, 88–9
Palaeontological Society, The, *296*
Palaeontology, Department of, 10–14, *15*, 29
Palmer, Phil, 264
parasites, 190
paratypes, 60
Parkinson, John: *Theatrum Botanicum, or Theater of Plantes*, 167
Parkinson, Sydney, 308, *310*
Partnerships for Enhancing Expertise on Taxonomy (PEET), 304–5
Parus montanus (marsh tit), 32
Parus palustris (willow tit), 32
Pasteur, Louis, 153, 189
Patterson, Colin (1933–98), 94–6, *95*
PCR (technique), 63, 141
Peake, John, 125, *126*, 137, 266
pegmatites, 242
Pelmatopora, *298*
Petiver, James (1658–1718), 158
petrology, 232–3
Phenacoccus manihoti (mealy bug), 199
Philosophical Transactions of the Royal Society, 308
PhyloCode, 70–1
phylogeny, 171, 289
Pillari, Professor (of Berne), 148–9
Piltdown Man hoax, 97–103, *101*, *102*
Planck, Max, 260
plants: collectors and herbaria, 157–8; hybridization, 311; naming, 166–8; pressing and preserving, 154–5, *155*, 157
Platnick, Norman, 176
Pleistocene period, 12
plesiomorphy, 94–5
plesiosaurs, 9, *10*
Plicolucina flabellata, *120*
Plot, Robert, 88*n*
Poe, Edgar Allan, 297
Polaszek, Andrew, 197
potoroos, 66, 68
Potter, Beatrix, 270
Press, Bob, 174
Prior, George: *Catalogue of Meteorites*, 223
Proust, Marcel, 192
proustite, 244
Prys-Jones, Robert, 278–9
pubic louse (*Phthirus pubis* "crabs"), 192
Punch (magazine), *198*, 260, *261*

Purves, Peter (1915–95), 146–51, *149*, 257, 285
Purvis, William, 161

quartz, 244–5

radiocarbon dating, 106
Ramsbottom, John, 159–60; *Mushrooms and Toadstools*, 159
rare earth elements, 227–8
Rasta thiophila (lucinid clam), *119*
Ray, John, 115*n*
Ray Society, 115
Rees-Jones, Deryn, 294
Reid, David, 114–16, 126, 303
reptiles, 151
Research Councils, 265, 302, 304
rhinoceros beetle, *214*
ribosomal ITS (Internal Transcribed Spacer), 68
Rio de Janeiro, Earth Summit (1992), 81*n*, 304
Rixon, Arthur, 108
RNA, 63–4
Roberts, Mark, 105
Roberts, Peter, 160*n*
Robson, Norman, 183–4
rocks, collection and study of, 230–3
Rollinson, David, 143
Ronson, Gerald, 122–3
Rose, Jonathan: *Intellectual Life of the British Working Classes*, 43
Rosen, Brian, 26
Ross, Robert, 178
Rothschild, Lionel Walter, 2nd Baron (1868–1937), 125, 195, *196*
Rothschild, Miriam (1908–2005), 194–6, 258
Rothschild, Nathaniel Charles (1877–1923), 195
Rothschild family, 194–5
Royal Entomological Society, 210, 216
Royal Scottish Museum, Edinburgh, 43
Royal Society, 309
Royal Society for the Protection of Birds, 125
Rudwick, Martin: *Bursting the Limits of Time*, 229*n*
Russell, Miles: *Piltdown Man: The Secret Life of Charles Dawson*, 102
Russell, Sir Arthur, Bt., 253–4, *254*, 255
Rutherford, Ernest, Baron, 295–6
Ryback, George, 280

Sahara Desert, 221–2
salmon, reintroduced to Thames, 136–7
Saunders, Ernest, 123
scanning electron microscope (SEM), *57*, 233–5
Schindler, Karolyn, 270
schistosome parasite (*Schistosoma*), 144–6, 302
Schmitz, B., 225–6
Schopf, Bill, 224
Science (journal), 78
Scott, Captain Robert Falcon, 232
Scott, Sir Peter, 128
Scott, Sir Terence Morrison, 127
Scottish Journal of Science, 58
screw worm (*Cochliomyia hominivorax*), *186*, 187–90, *188*, *191*
Scrope, George (1797–1876), 308
scuttle flies (Phoridae), 304
sea spiders (pycnogonids), 142
sea urchins (echinoids), 176
Second World War (1939–45), 283–5, *283*
sequoia tree, 23
sexton beetle (*Nicrophorus orbicollis*), 203
Shaw, George, 158
Shaw, George Bernard, 295
shells (molluscs), 19, 44, *117*
Sherard, William, 167
Shipley, A. E.: *The Minor Horrors of War*, 190*n*
Siebert, Darrell, 135
silicates, 228
Siveter, David, 53
Siveter, Derek, 142
6323 (house magazine), 285
slime moulds, 217–18
Sloane, Sir Hans (1660–1753): bequeaths collection to Trustees of British Museum, *21*, 120; glass models of jewels, 251; herbarium, 157–8; *Hortus siccus* (dry garden), 158; mineral collection, 255; monument, *121*; overseas visitors to collection, 167; voyage to Jamaica (1687), 157, 164; zoological specimens decay, 158
Smirke, Sir Robert, 158

Smith, Andrew, 176
Smith, Campbell, 240
Smith, Ken, 202
Smith, Sir James Edward, 169
Smith, William (1769–1839), 87–8
Smithsonian Institution, Washington, National Museum of Natural History, 31–2, 313
snails (Gastropoda), 114–17; as hosts to *Schistosoma,* 144–5
SNC meteorites, *see* meteorites, from Mars
Snow, Charles P., Baron, 160
Solanum (genus), 168, 170–3, 175
Solanum cheesmaniae, 272
Southgate, Vaughan, 143
Spath, L. F. (1882–1957), 88–90, 93, 94, 112, 299*n*
species: classification and identification, 32–3, 36, 58–61, 68–71, 305, 311–12; extinction threats and conservation projects, 288–9; new discoveries, 32–5, 59–60; and variants, 311
specimen labels, 13–14
spiders, internet guide to, 176
Spigelius, Adrianus (1578–1625): *Isagoges in rem herbarium,* 155
Spirit Building, 26, 29
Spitalfields Project, 275
Spooner, Brian, 160*n*
springtails (Order Collembola), 208–9, *209*
Stammwitz, Percy, 276
Stammwitz, Stuart, 276
Stanley, Chris, 230
staphylinids (rove beetles), 215
Staunton, Sir George, 284
Stearn, William T., 3, 46, 138, 258, 281
sterilization, as biological control, 189–90
Stork, Nigel, 34
Story-Maskelyne, Nevil Mervyn Herbert (1823–1911), 251
Stott, Rebecca, 91
stratigraphy, 86–8, 90, 93–4
Stringer, Christopher, 104–5
Sulawesi, insect inventory, 216
Sullivan, Admiral J., 259
Sutcliffe, Tony, 101, 258
Sweden, limestone quarries, 225–6
Sydney, Australia, Botanical Garden, 47
synapomorphy, 94–5
Systematics, 32, 35, 96–7, 177, 289

Tabernaemontanus (Jacob Theodore von Bergzabern), 157
Tahiri, Ibrahim, 221–2
Talling, Richard, 255
Tams, "Tiger" (beetle expert), 290
Tandy, Geoffrey, 165–6
Tate Regan, Charles (1878–1943; Director 1927–38), 134, 262*n,* 273
taxonomy: alpha and beta, 304; and DNA Barcode, 306; funding, 303–5; and genome, 63; and holotype, 60; importance, 304; "lumpers" and "splitters," 311; and nomenclature, 48–51, 52–4; and revision, 68–9; role and purpose, 32–3, 62, 289; *see also* Linné, Carl von
Taylor, John D., 19–20, 117–18, *119*
Taylor, Paul, 235, 298
Teilhard de Chardin, Marie-Joseph, 99, 103
Tennessee, University of, *see* University of Tennessee
termites, 210–11
Thames, River, cleaned up, 136–7
Thames Estuary Partnership, 136
Thames Salmon Rehabilitation scheme, 136–7
Thames Salmon Trust, 136
Thatcher, Margaret, Baroness, 122
Theobroma cacao, 157
Thomas, Oldfield, 23
Thompson, Sir Edward Maunde, 260
Thomson, James: "The Seasons," 166
Tierra del Fuego, 179
Tilley, Cecil, 233
Times, The, 47, 262
Tin Hat (wartime house magazine), 285, *286*
tomato family, *see Solanum*
transmission electron microscope, 235
Treatise on Invertebrate Paleontology, The (University of Kansas series), 302–3
Tree of Life project, 304
Trewavas, Ethelwynn (1908–93), 134–5
trilobites: characteristics, 73–4, *76,* 78–9, *78, 79;* diversity, 74, 308; in High Atlas, Morocco, 74–9; pronunciation, 73; under SEM, 235; study of, 8–9; in

Swedish limestone, 225–6; in *The Treatise*, 302
Tring, Hertfordshire, 125, 195
truffles, 65–9, *65*
Tucker, Denys W., 126–8, 135
types and type specimens, 60–2

Ulrich, E. O., 31
Underwood, Garth, 151
Universidad Nacional Autónoma de México, 173
University of Tennessee, Knoxville, Department of Anthropology, 204
Uppsala, Sweden, 49, 50, 54
Ural mountain chain, 238, 239

Vane-Wright, Dick, 200, *201*, 206, 216, 288–9
vent faunas, 238–9
Victoria, Lake (Africa), 133–4, 135
Victoria, Queen, 38, 250
virus vectors, 198
volcanogenic massive sulphides (VMS), 236–9

Wall, Frances, 240–1, 242
Wallace, Alfred Russel, 170–1
Walliser, Otto, 74
Wandsworth, south London, Museum acquires premises in, 122–3
Wanless, Fred, 125
Warlow, Andy, 141–3
wasps, parasitic, 197–9
Waterhouse, Alfred, designs Natural History Museum building, 38, 293
Way, Kathy, 18, 19
Weiditz, Hans, 155
Wellcome Foundation, 146
Wells, H. G., 192
Wernjam, Herbert, 178
whales, 146–8, *149*
Wheeler, Alwyne (1929–2005), 136–7
Wheeler, Quentin, 218–19, *218*
Whitaker, Amoret, 202
White, Rev. Gilbert, 286, 309
white flies, 197–8
Whitehead, Peter J. P.: antagonism with Wheeler, 136–8; character and career, 129–33, *130*, *132*, 135, 143; *Clupeoid Fishes of the World*, 128; on *Horizon* programme, 257–8
Whitehead, Sir Rowland, 129
Whittington, Harry, 302
Whybrow, Peter, 82
Williams, David, 180–2
Williams, Paul, 289
Williams, Shirley, Baroness, 29
Williams, S. T., 116
Williams, Terry, 233
Wilson, E. O., 135, 207, 306
Winchester, Simon: *The Map That Changed the World*, 87
winkles, 114–17, *117*
Winthrop, John, 255
Withers, T. H.: *Catalogue of Fossil Cirripedes* (1928–53), 90, 93
Wolfson Foundation, 146
women, in Museum, 269–75
Wooddisse, Thomas, 263
Woodward, Arthur Smith, 97–9, 100, 103, 270
Woodward, Henry, 270
Woolley, Alan, 240–1
Wordsworth, William, 244–5
World Health Organisation, 144
Worldmap Project, 289
World Wars, *see* First World War; Second World War
Worms, Baron de, 210, 266
Wright, Chris, 143
Wyatt, Colin, 206

X-ray CT (computed tomography) technology, 83

Yaman Kasy, 238–9
Yarrow, Ian, 200
Yttrium, 227–8, 242

Zacharias, J. F., 187*n*
Zguilma (hill), Morocco, 75, *76*, 77
Zinnwaldite, 230
zoological ranks, 50*n*
Zoology Department, 26–7, 29

ALSO BY RICHARD FORTEY

"Give me a scientist with a sense of humor and a knack for metaphor who also happens to have an encyclopedic understanding of natural history, and I will always be happy. Mr. Fortey is such a writer."
—Richard Bernstein, *The New York Times*

EARTH
An Intimate History

Beginning with Mt. Vesuvius, whose eruption in Roman times helped spark the science of geology, and ending in a lab in the West of England where mathematical models and lab experiments replace direct observation, Richard Fortey tells us what the present says about ancient geologic processes. He shows how plate tectonics came to rule the geophysical landscape and how the evidence is written in the hills and in the stones. And in the process, he takes us on a wonderful journey around the globe to visit some of the most fascinating and intriguing spots on the planet.

Science/Evolution/978-0-375-70620-2

LIFE
A Natural History of the First Four Billion Years of Life on Earth

From its origins on the still-forming planet to the recent emergence of *Homo sapiens*—one of the world's leading paleontologists offers an absorbing account of how and why life on earth developed as it did. Interlacing the tale of his own adventures in the field with vivid descriptions of creatures who emerged and disappeared in the long march of geologic time, Richard Fortey sheds light upon a fascinating array of evolutionary wonders, mysteries, and debates. Brimming with wit, literary style, and the joy of discovery, this is an indispensable book that will delight the general reader and the scientist alike.

Science/Biology/978-0-375-70261-7

TRILOBITE
Eyewitness to Evolution

They lived in the oceans over five hundred million years ago, and they were as myriad then as the beetle is now. Their eyes were made of calcite lenses that could number from one to a thousand. Some were spiky while others were smooth; some were larger than lobsters while others were smaller than fleas. They were trilobites, crustacean-like animals whose span on Earth was so long—three hundred million years—that to study them is to glimpse a different world ancient continents now positioned elsewhere and mighty oceans long since vanished. Richard Fortey has been enraptured by trilobites since the age of fourteen, and his enthusiasm is exuberantly evident on every page of this book. *Trilobite* is that rare volume as compelling for the layman as the specialist, and an elegant testimony to the joys of scientific discovery.

Science/Evolution/978-0-375-70621-9